Earth: The Sequel

Earth:
The Sequel

THE RACE TO REINVENT ENERGY
AND STOP GLOBAL WARMING

Fred Krupp and Miriam Horn

W. W. NORTON & COMPANY

New York London

For information about permission to reproduce selections from this book,
write to Permissions, W. W. Norton & Company, Inc.,
500 Fifth Avenue, New York, NY 10110

For information about special discounts for bulk purchases, please contact
W. W. Norton Special Sales at specialsales@wwnorton.com or 800-233-4830

Manufacturing by RR Donnelley, Harrisonburg
Production manager: Julia Druskin

Library of Congress Cataloging-in-Publication Data

Krupp, Fred.
Earth, the sequel : the race to reinvent energy and stop global
warming / Fred Krupp and Miriam Horn.
p. cm.
Includes index.
ISBN 978-0-393-06690-6 (hardcover)
1. Energy conservation—United States. 2. Global warming. I. Horn,
Miriam. II. Title.
TJ163.4.U6K78 2008
621.042—dc22

2008001317

W. W. Norton & Company, Inc.
500 Fifth Avenue, New York, N.Y. 10110
www.wwnorton.com

W. W. Norton & Company Ltd.
Castle House, 75/76 Wells Street, London W1T 3QT

2 3 4 5 6 7 8 9 0

FOR ALEX, ZACH, AND JACKSON KRUPP,
FOR FRANCESCA SABEL,
AND FOR EVERYONE ELSE'S CHILDREN AS WELL

CONTENTS

Earth: The Sequel

CHAPTER 1

A New Industrial Revolution

A revolution is on the horizon: a wholesale transformation of the world economy and the way people live. This revolution will depend on industrial technology—capital-intensive, shovel-in-the-ground industries—and will almost certainly create the great fortunes of the twenty-first century. But this new industrial revolution holds a more important promise: securing the world against the dangers of global warming. It is developing amid the political, economic, and technological equivalent of the perfect storm: worldwide concern about the enormous threat of greenhouse gases, growing realization that we are prisoners of petroleum—hostage to the unstable, sometimes hostile, regimes that control the supplies of crude oil and natural gas—and, finally, huge and accelerating advances in technology that make possible unprecedented breakthroughs in how we make and use energy.

This book is about the kinds of inventors who will stabilize our climate, generate enormous economic growth, and save the planet. It is also about the near certainty that unless the United States acts as a nation to give these innovators the chance to compete fairly in the world's biggest business, they will fail to avert the crisis in time.

————

A BIT OF CLEAN-AIR HISTORY will explain how U.S. leadership can solve the most dire environmental problem of our times—how collectively we can head off catastrophe, create wealth and jobs, and enhance our security.

Thirty years ago, scientists and fishermen began noticing a startling decline in animal and plant life in lakes and forests throughout the eastern United States. In 1981 the National Academy of Sciences issued a report documenting the cause: sulfur dioxide (SO_2) pollution. Originating mostly from coal-fired power plants, sulfur dioxide was causing rain and snow to turn acidic, and that acid rain was killing aquatic life and damaging forests.

The discovery sparked bitter debate over how to reduce sulfur dioxide emissions and at what cost. The key instrument then available for tackling acid rain was the 1977 Clean Air Act, which had established a "command and control" approach to air pollution. Not only was every newly built electric generating unit required to meet a uniform ceiling on its emissions rate, but it had to do so by installing a "scrubber," or flue-gas desulfurization unit, an expensive piece of capital equipment that added tens of millions of dollars to the cost of the power plant. It didn't matter if a power plant could cut emissions more cheaply by burning cleaner coal or by making its boiler more efficient—it still had to install a scrubber. Meanwhile, overall emissions continued to rise: old plants were largely exempted until they modernized, and new plants, even if in compliance with the law, still added new emissions to the total.

The one-size-fits-all solution failed to take advantage of the local knowledge and experience of plant operators. It actually impeded quick adoption of new technologies. And it made the whole process far more expensive than necessary: some existing plants could cut pollution cheaply but weren't asked to, while other plants spent a fortune to comply.

By the 1980s, Environmental Defense Fund had taken the lead in arguing for a different approach. The key was to mobilize the market: to make far deeper cuts in pollution and at much lower cost. During the 1988 presidential election, strong pressure by New Hampshire environmentalists elevated acid rain as an issue all candidates had to address. Soon after George H. W. Bush was elected president, Environmental Defense Fund president Fred Krupp met with White House counsel C. Boyden Gray. He outlined for Gray how the president's campaign pledge to curb acid rain could be fulfilled using the world's first "emissions cap and trading system." He argued that the president could be more ambitious on the environmental goals while still garnering business support. After dozens of meetings, the Bush administration embraced Environmental Defense Fund's proposal and submitted it to Congress, which—led by Senator George Mitchell of Maine and Representatives Henry Waxman of California and John Dingell of Michigan—wrote it into the Clean Air Act of 1990.

The law started with the scientific bottom line. It required a 50 percent cut in sulfur dioxide emissions from the total volume released by fossil fuel–fired power plants—the minimum reduction that atmospheric scientists believed was necessary to begin bringing lakes and rivers back to life. It set a permanent upper limit—a cap—on these emissions, then divvied up the quantity (in tons) of pollution allowed among the power plants. Even as new sources of pollution came on line, the cap was ratcheted down over time so that the total amount of pollution fell. That declining cap, guaranteeing that environmental targets would be met, was unprecedented. But it was the second part of the law— the emissions-trading system—that completely transformed the paradigm that had historically pitted environmentalism against economic growth.

The trading mechanism allowed a power plant that cut its sulfur dioxide pollution more than required to sell those extra allowances, and permitted plants that could not find a better way to cut their own emissions to buy them. A new commodities market was born. A plant that could beat its emissions target had a profitable new asset to sell, and financial incentive to develop ways to cut emissions even further. The buyer had the flexibility to find the cheapest way to meet the cap; it was now the power plant operator—not the regulator—who decided how to integrate emissions control into an overall business plan.

Two months after the law was passed, Richard Clark, then CEO of Pacific Gas and Electric, America's largest publicly owned utility, sat next to Fred Krupp at a dinner for the President's Commission on Environmental Quality. "When you were talking to the president about this cap-and-trade idea, I frankly thought you'd lost it," he said to Fred. "But now that there's a way to make money from cutting pollution, I have a dozen proposals for emissions reductions from my own employees on the shop floor, and a dozen more from outside consultants. The environment isn't just a money loser—it's a profit center. I have to admit it's a powerful law."

One company that jumped on the new market opportunity was General Electric (GE), whose scrubber technology—in the absence of strict caps on emissions or any rewards for overachievers—had not advanced much since passage of the 1977 law. For years, GE had been selling clumsy units that clogged up so frequently that operators had to build two parallel scrubbers to make sure one was always running. Eli Gal, whose current carbon-cutting work is featured in Chapter 8, was then at GE Environmental Services, and he recalls how quickly that all changed once the cap-and-trade mechanism became law. GE began devoting serious resources to clean-up technology, and Gal's team had its breakthrough: it devised a scrubber that turned the sulfur dioxide into gypsum, which does not gum up the works and is itself a marketable product.

Elsewhere, the new law inspired innovative thinking about low-tech solutions. For decades, conventional wisdom held that low-sulfur coal from Wyoming could not be mixed with high-sulfur coal from Appalachia—at least, not in quantities large enough to significantly reduce sulfur dioxide emissions. The new incentives made it worthwhile to take a second look at that belief—and before long, scientists and engineers proved the conventional wisdom wrong. Capital and ingenuity were suddenly flowing to solutions both high tech and low, for there was profit to be made.

Within five years, U.S. utilities cut emissions 30 percent more than the law required, even while increasing electric generation from coal by 6.8 percent and reducing retail electricity prices. During that same period, the U.S. economy grew by a healthy 5.4 percent. Dire predictions that the program would eventually cost more than $6 billion a year proved wildly off base; recent studies peg the actual cost at between $1.1 and $1.8 billion. And by 2000, scientists were documenting decreased sulfates in Adirondack lakes, improved visibility in national parks, and widespread benefits to human health.*

THE CAP-AND-TRADE MECHANISM had unleashed the most powerful economic force in the world in the service of environmental goals: entrepreneurial capitalism. The chapters that follow demonstrate that the same market forces can be brought to bear on the problem of global warming.

---

* By the time the program is fully carried out in 2010, it will provide more than $100 billion in annual health and environmental benefits, preventing numerous deaths, hospitalizations, and emergency room visits while restoring the vitality of natural systems.

As well as markets have worked over the years—spurring the production of electronics of higher and higher quality at lower and lower prices, for example—they have failed the environment because they have failed to account for the cost of pollution. No cost attaches to the global warming pollution emitted by power plants, factories, and cars, and no benefit flows to those who reduce that pollution. The legendary venture capitalist John Doerr, an early funder of Google and Amazon.com (and a supporter of Environmental Defense Fund; Ann Doerr, his wife, is on the board of trustees), puts it this way: "Every single day, we dump 70 million tons of carbon dioxide into our atmosphere like it's an open sewer, like it's entirely free to do that. It's really hard to change consumer behavior when consumers don't know how much their behavior costs."

The solution is, at its core, simple. The U.S. Congress must set a legal and steadily declining limit on global warming pollution. The allowances will be divvied up among emitters, or auctioned by the government to raise revenues—or some combination of the two. Polluters who emit more will need to pay for the extra pollution reductions achieved by others; those who can reduce global warming pollution further will profit by selling those reductions in an open market. We can, in short, use the power of the market system to climb out of the hole created by flawed markets. We can offer a pot of gold to those who develop new ways to generate carbon-free energy and new technologies to remove carbon from our smokestacks and atmosphere. We can channel the full range of human impulses—ingenuity, idealism, ambition—into undoing the damage and healing our planet. America's greatest strength has always been its boundless capacity for invention. In the words of one energy entrepreneur, "The beauty of this country is that every time we're pressed to the wall we come up with new things—we become the most creative force in the world."

GIVEN ENOUGH TIME, many of the technologies described in this book might ripen on their own. But time is the one thing we don't have. Late in 2007, scientists noted that emissions of greenhouse gases were growing considerably faster than expected.* They also dramatically revised, downward, their estimates of how much pollution the planet can tolerate. The new numbers mean that we must not simply stop the annual growth of emissions. *Worldwide, nations must cut emissions in half over the next fifty years. To reach that goal, the United States will have to cut emissions by 80 percent.*

The scientific consensus is that inaction will change the earth within a few decades into a place unlike any ever inhabited by humans. Business as usual will open the door to catastrophe: flooding and the dislocation of millions of people in South Asia's vast deltas; chronic drought and mass malnourishment in Africa; wildfires, deadly heat waves, and coastal destruction in the United States; the extinction of half the world's living species. U.S. generals warn of unprecedented waves of refugees, and of threats to national and global security from wars over resources. In the context of such all-too-plausible scenarios, the wholesale reinvention of the way we make and use energy is not a choice but a matter of survival.

Securing our planet against calamity will require a second industrial revolution as sweeping as that effected a century ago by the likes of Thomas Edison, Henry Ford, and John D. Rockefeller. We will need to harness energy from the sun, the waves, living organisms, and the heat embedded in the planet. We will need to reinvent automobiles, clean up emissions from the immense and rapidly

---

* Stephen Pacala and Robert Socolow of Princeton University's Carbon Mitigation Initiative had projected emissions of 7 billion tons of carbon for the year 2004; instead, emissions for that year surpassed 8 billion tons. Current trends suggest emissions are outpacing even the worst scenarios outlined by the Intergovernmental Panel on Climate Change.

growing coal infrastructure, use the energy we have far more efficiently and put an end to tropical deforestation. What will become apparent on these pages is that we can achieve this revolution, but only if we make it a fair game. No single technology will stop global warming, but there is a silver bullet: a cap on carbon that will launch all these solutions into the mainstream.

THE INNOVATORS YOU WILL MEET in the chapters that follow are wildly inventive and ambitious. Risk-taking is their favorite sport, "disruptive" their favorite word. Many actually relish the prospect of going up against the biggest companies in the world. As an undergraduate at MIT, Neil Renninger—now a chemical engineer at Amyris Biotechnologies—supplemented his income by playing on that university's legendary blackjack team, whose exploits were chronicled in the 2002 book *Bringing Down the House*. He faces Amyris's challenges with the same spirit he brought to the casinos. "If you take educated risks, figure out where you're advantaged, and play to those advantages," he says, "you can beat the house."

The investors who fund these inventors are just as certain of their power to make historic changes. Silicon Valley leaders like John Doerr and Vinod Khosla, a cofounder of Sun Microsystems who is now also a major venture capitalist, have shifted immense resources into clean-energy technologies (Doerr calls it the "mother of all markets"). Where others look at the trivial market penetration of, say, solar technology and see evidence that alternative energy will remain marginal, these investors see a massive market opportunity: to meet China's stated goal to derive 10 percent of its electricity from renewable sources (not counting large hydroelectric projects) by the year 2010 will require 6 gigawatts of electricity—more than two years of output from all the solar-cell factories in the world today.

Yet they also know that whether they succeed or fail depends ultimately on what we demand from our political leaders. The current rules of the game are steeply stacked against the new-energy entrepreneurs. Even the best ideas will fail in a contest as rigged as the $5 trillion energy business is today. While innovators developing semiconductors and the Internet got to play in a virtually open field, these energy innovators are up against the most powerful companies in the world, companies that have spent decades successfully pushing for subsidies, trade agreements, and regulatory structures that favor their business. Oil and gas companies spend some $60 million a year lobbying for policies that favor their industries and receive, according to the U.S. Government Accountability Office, benefits worth $6 billion a year. Incumbent companies control pipelines and transmission grids; the high cost of upgrading and connecting to the grid can strangle a start-up renewable-energy plant. Venture capitalist Doerr observes that each year the federal government devotes just $1 billion to research on renewable sources of energy—less than ExxonMobil earns in a single day. Most important, policymakers are only just beginning to confront the huge hidden subsidy for fossil fuels: that no financial account is taken of the use of the atmosphere as a dumping ground for the pollutants that cause global warming.

Though America's representatives have been aware of global warming since 1988, when Senator Al Gore held the first congressional hearings on the data then beginning to emerge, only in recent months has the U.S. Congress displayed real interest in addressing climate change. Some of the legislative proposals are hollow grandstanding; others would ensure meaningful action. There are many bad ways of moving forward from here, including continuing to give taxpayers' money to the businesses with the best lobbyists. The huge federal subsidies for corn ethanol, for instance, are chiefly testament to the power of the agribusiness giant Archer Daniels Midland and other agricultural interests.

The history of the development of wind and solar power—booming in times of high oil prices and subsidies, busting when those disappear—reinforces the necessity for long-term policies that do not require Washington to pick winners, but that allow the United States' best handicapper—the market, freed of existing flaws—to sort out who really can deliver the goods. Much of the U.S. business community has already embraced that view. When the California legislature was debating passage of the nation's first economy-wide cap on carbon emissions, clean-tech proponents effectively countered claims that carbon legislation would devastate the economy, by detailing the immense opportunities for job growth, profits, and international competitiveness that would be unleashed by the proper regulatory framework. They reminded lawmakers that technological progress driven by innovation is responsible for fully half the growth of the U.S. economy. They found allies in a number of the country's workhorse companies, which noted that *not* taking action would almost surely have dire consequences for their business.

In the spring of 2007 ten leading U.S. companies, including General Electric, Alcoa, Caterpillar, and Duke Energy, called for a national cap on carbon emissions; in the months that followed they were joined by dozens of Fortune 500 companies, including the big three U.S. automakers and Shell Oil. And as California and states in the Northeast and Northwest moved ahead with caps on carbon emissions, the nation's most innovative companies began operating as if a federal carbon cap were already in place. As it switched on a 1.6-megawatt solar array at its Silicon Valley headquarters—the first of 1,000 megawatts of renewable energy it ultimately plans to produce—Google announced that it would "set an internal cost of carbon voluntarily by using a 'shadow price,' the theoretical cost of carbon that we expect under a regulatory market. This will allow us to make operational decisions as if there were already a price on carbon. That in turn enables us to include the true cost of power as one of the key criteria in site selection for our data centers—a cost not yet being recognized by the market, but one that will soon become real through carbon legislation. This is an important tool to reduce the financial risk that our energy investments face, and when evaluating power options, it will also put renewable energy on a level playing field."

A level playing field. That is what these venture capitalists and innovators require—not assurances of profits but enough of a fighting chance to make the huge risks they are taking reasonable. Many still will not succeed. Those who do will need enormous ingenuity and doggedness to make their inventions work at a marketable price. But while you read about these inventors and get a glimpse of what is possible, imagine what it will mean if even a fraction of these high-risk ventures fulfill their promise—if clean, abundant energy becomes a reality.

The stakes are almost unimaginably high.

CHAPTER 2

Harnessing the Sun, Part I

I n 2005 the world's solar energy–generating capacity grew by 44 percent. If that pace can be sustained over the next few decades, by 2050 the sun could supply ten times as much energy as the earth needs. Such a growth rate might seem like wishful thinking. But it is worth remembering that the semiconductor industry has grown at an even faster rate for a similar length of time. Innovators and investors took the personal computer industry from zero machines to almost a billion in thirty years, doubling processing speeds every twenty-four months or less and cutting costs in half each time the speed doubled. As Oliver Morton wrote in the September 7, 2006, issue of the science journal *Nature*, "If Silicon Valley can apply Moore's law★ to the capture of sunshine, it could change the world again."

To change the world again, and to get very rich doing so—that's why venture capitalists are pouring billions into start-ups developing photovoltaic cells, which convert sunlight directly to electricity. (The other way to use the sun to make power is by tapping its heat, a strategy explored in Chapter 3.) For investors who made their

---

★ See Intel.com: "In 1965, Intel co-founder Gordon Moore saw the future. His prediction, now popularly known as Moore's Law, states that the number of transistors on a chip doubles about every two years."

first fortunes from semiconductors and the Internet, the learning curve on photovoltaics is not terribly steep. Solar power has grown up alongside the chip industry, borrowing its materials and processes and, increasingly, its talent. The geographies of the two industries overlap. Many of the solar start-ups are in California's Silicon Valley, in Cambridge, Massachusetts, in Phoenix, Arizona, and in Austin, Texas. And many have close relations with the same universities: Stanford; University of California, Berkeley; the California Institute of Technology; and MIT.

For those who believe that individuals around the world have gained more control of their destiny through the computing and networking revolution, solar power has a further appeal. Because photovoltaic cells produce electricity where they are used, they have the potential to reshape the centralized energy economy into something more like the network created by the advent of personal computers, cell phones, and the Internet. Owners of photovoltaic systems can become power producers themselves, sell energy into the grid at a profit, even gain near-term access to electricity in parts of the world that do not yet have energy infrastructure; already, photovoltaics are bringing electricity to poor villages in Africa for about $250 per household. "Distributed energy" erases the strategic advantage of big energy companies, says Andrew Beebe, president of Energy Innovations.* "With solar, they can't control the resource. That's a real shift of power."

EVERY HOUR, the sun provides the earth with as much energy as all of human civilization uses in an entire year. At just 10 percent efficiency—that is, if only 10 percent of that solar energy were converted to electricity—a square of land 100 miles on a side could produce enough electricity to power the entire United States. Those

---

* From a January 22, 2007, interview with Worldchanging.com.

two facts would seem sufficient in themselves to map the solution to global warming. Yet a century after Albert Einstein explained the photoelectric effect (for which he won a Nobel Prize), and fifty years after Bell Labs invented the first semiconductor-based device to convert sunlight into electricity, solar technology remains a trivial player in global energy. In 2007 the total solar capacity worldwide was just 6.6 gigawatts, compared to more than 1,000 gigawatts for coal; in the United States, solar cells provided less than 0.05 percent of the electricity supply.

Part of the reason is the complexity of that supply. Behind every electrical outlet there is a vast web of resources, and introducing new technologies into that web has cascading and sometimes counterintuitive impacts. Still more important is the nature of sunlight itself. The sun does not shine twenty-four hours a day or every day of the year—far from it in some regions. And that creates the need for cost-effective storage: some way to capture and save that intermittent energy so that it is available on demand, even when the sun is not shining. The same problem confronts other renewable energy sources as well—especially wind.

Traditionally, energy storage has meant batteries. But while batteries can inexpensively provide short bursts of power, using them to store the large volumes of energy needed when the wind stops or the sun doesn't shine, possibly for many hours or days, is prohibitively expensive. Researchers are therefore working not only to improve batteries but also to develop alternative storage technologies—using excess electricity to pump water up into reservoirs for use later in hydroelectric generators, for instance. Until such technologies are greatly improved, however, storage will remain a major impediment to widespread use of solar energy.

The biggest obstacle is that photovoltaics are not yet cheap enough to compete at scale. Three paths could potentially get them there. The first is to continue stepping up efficiencies of existing technologies, primarily crystalline-silicon cells, while lowering costs. The second is to leap to cheap next-generation technologies that can be produced in quantity: making as many square miles of photovoltaic foil or fabric (or paint or Astroturf) as possible, even if it generates less energy per square foot. The third is to pay a price premium for quality: cramming the most efficiency onto the smallest possible cells, then wrapping those cells in optics that concentrate the sunlight, multiplying its intensity five hundred or one thousand times.

To make sense of these paths, it helps to understand the basic principles of photovoltaics. Solar cells are made of semiconducting elements, most commonly silicon, which hold on to their electrons until hit with the necessary oomph, or "band-gap energy." When a photon bearing light energy from the sun strikes with that threshold amount of energy, it can kick an electron free from the silicon atom and up to the conduction band, leaving a vacancy known as a hole. Each electron-hole pair, bearing opposite charges, is called an exciton. The electrons flow through the conduction band to reunite with the holes. That's an electric current.

It also helps to understand how the energy produced by photovoltaics is priced. Though electricity is usually priced by the kilowatt-hour,* solar manufacturers talk about the "price per peak watt," referring to the maximum output of their cells during peak sunlight. In 2007 the peak watt price averaged about $4 (or just under $7 installed—a price that includes the cost of all hardware, mounting and electrical systems, engineering, and installation). The manufacturers are not just being difficult, but rather are trying to

---

* A brief primer on watts: A watt (W) is the standard unit of electrical power. One kilowatt (kW) equals 1,000 watts, or enough power to light ten 100-watt light bulbs at any given moment. If you want to light those ten bulbs for an hour, you will need one kilowatt-hour, which is the unit used on most electricity bills. One megawatt (MW) equals 1,000 kilowatts. A gigawatt (GW) equals 1,000 megawatts, and 1 terawatt (TW) equals 1 million megawatts.

take account of the fact that the sun shines more and brighter in some places; in those places the very same system will produce a lot more energy and will therefore cost less per kilowatt-hour.

Take a typical single-family rooftop installation. For a system capable of generating 3 kilowatts, you would currently pay about $21,000. And in an hour of peak sunlight, you would get 3 kilowatt-hours. If you lived in Las Vegas, however, you would get many more of those peak sunlight hours than if you lived in, say, Fairbanks, Alaska. In fact, the yearly output of your 3-kilowatt system in Vegas would be about 6,500 kilowatt-hours, nearly twice the 3,300 kilowatt-hours you would get in Fairbanks. If your rooftop panels lasted thirty years, your price per kilowatt-hour would be 11 cents in Vegas, 21 cents in Fairbanks. That might be cheaper than the electricity you buy from your utility, but again that depends on where you live. Analysts for the French brokerage Crédit Agricole estimate that in Tokyo, where retail electricity prices are extremely high and there is moderate sunshine, "solar is cost competitive at $5 a watt [more than $8 installed]. Los Angeles is close behind (even more sunshine; nearly as costly energy), while solar will not become cost competitive in Portland [Oregon] any time soon."

Though solar power is generally judged on the basis of whether it can beat the retail price of coal-generated electricity, the comparison misses a key point. The greatest value of solar power is that it is most productive when the weather is sunny and hot—precisely when consumer demand forces a utility to operate at full throttle. For PG&E, for instance, peak demand is growing 25 percent faster than overall electricity needs, but the last quarter of capacity is needed less than 10 percent of the time. "Peak shaving" with electricity produced by rooftop solar panels relieves that pressure on utilities, damping the prime driver to build new gas-fired plants. It also reduces costs. Utilities "turn on" power plants in order of their variable cost, starting with the least expensive plant to operate and moving to the most expensive as demand rises. Solar installations produce electricity at no extra cost just when units with very high variable costs would otherwise be called upon to run, providing tremendous savings for the utility, the consumer, and the atmosphere. (To date, most owners of solar photovoltaic systems do not get the discount they deserve on their electric bills. Typically, utilities charge customers the average cost of generating electricity over the course of a day. Some are now experimenting with "real-time pricing," charging customers hour by hour for the true cost of the electricity they use.)

Given all that, almost everyone in the industry agrees that when the price per peak watt falls to $1 and the storage problem is solved, solar-generated electricity will compete with coal-fired electricity virtually everywhere.

Producers of traditional crystalline-silicon solar cells, which have dominated the market for thirty years and in 2007 still controlled 93 percent of it, believe they will get to that price with the help of the world's low-cost manufacturers. In 2007 China became the third biggest producer of solar cells, behind Japan and Germany, and raised billions from public stock offerings to expand capacity further still, creating, along the way, several new billionaires. Its leading company, Suntech, is worth $5.5 billion, employs thirty-five hundred people, and sells 90 percent of its output to Germany. Kyocera Solar, the Arizona unit of Japan's Kyocera Group, plans to build a factory in Mexico able to produce 150 megawatts of solar cells annually, enough to put new 3-kilowatt systems on fifty thousand homes. Japan's Sharp Corporation, the biggest solar-cell maker in the world, intends to increase its production capacity, all by itself, to 100 gigawatts by 2030.

India, where the government pays up to the full cost of solar projects in nonelectrified rural areas, is also an emerging powerhouse. BP has a solar joint venture with Tata in Bangalore, and

Moser Baer India has contracted with Applied Materials to build a solar factory in New Delhi by 2009. Hong Kong and Taiwan are increasingly important players as well. In addition to manufacturing crystalline-silicon cells, some Asian companies are also making lower-efficiency thin films on glass from amorphous (noncrystalline) silicon, employing a technology similar to that used in manufacturing flat-screen liquid crystal displays.

Though this global expansion has caused a shortage of crystalline silicone, industry research firm Clean Edge predicts that revenues in the solar photovoltaic industry will grow to $50 billion a year by 2015, reaching a total installed base of 75 gigawatts, a tenfold increase from today. But that would still supply just 0.5 percent of the total amount of electricity needed for 2015. A more rapid expansion will almost certainly require the next-generation photovoltaic technologies now emerging from labs into the commercial market.

Crystalline silicon has limitations. For example, it absorbs light slowly, meaning that wafers have to be thick (and heavy and expensive) enough to capture photons before they slip through. That thickness, in turn, requires highly purified silicon, with no more than one impurity per trillion parts.*

Such limitations have led most next-generation solar-cell makers to abandon silicon in favor of other semiconductor materials, usually several in combination: by mixing or stacking elements with different band gaps (threshold energies), they can harvest a wider range of wavelengths of light, wasting less incoming solar energy. Many have also abandoned wafers in favor of "thin films," replacing expensive batch processing and heavy, unwieldy modules with cheap, fast, roll-to-roll manufacturing of acres of flexible materials

---

* An energized electron deep inside has to stay energized all the way to the top; if the electron encounters defects or traps along the way, the energy degrades into heat before the electron completes its journey into the electrical circuit.

that can go anywhere. Others are developing solar concentrators, which combine small, ultra-high-efficiency cells with low-cost optics. Like the magnifying glass that scorches the hapless ant, these make diffuse solar energy hundreds of times more powerful.

INNOVALIGHT, AN EARLY-STAGE COMPANY based in Santa Clara, California, is making thin films but is one of the rare cutting-edge solar companies still using silicon. CEO Conrad Burke believes it is the only material benign and abundant enough (it makes up 15 percent of the earth's crust) to supply the staggering amount of electric power the world needs. "God put silicon on this planet for a reason," he says. "Not really, that's just me being Irish and Roman Catholic. I don't think He figures in it. But when those guys at AT&T built the first transistor with silicon, well you know what that set off. Fast-forward ten years . . . and I'm telling you silicon will win this war." What Burke has in mind is not, however, conventional silicon wafers. Innovalight has instead made something brand new, using nanotechnology—the engineering of materials at the atomic scale—to overcome nearly all the limitations silicon has in bulk form.

While the purified silicon used for both microchips and solar cells has been plagued with shortages and price spikes, for instance, Innovalight has found a way to bypass that supply chain, making its nanosilicon powder from cheap, unpurified sources. "People have figured out how to make micrograms of nanosilicon in days," says Burke, standing outside the closed doors of the labs where the closely guarded work is being done. "We can scale to kilograms." Innovalight also reports that it has reduced the amount of silicon needed per watt from the 15 grams for a conventional solar cell to just 0.04 gram.

Nanotechnology has also made possible high-throughput manufacturing, which Innovalight expects to cut costs by a factor of ten compared to growing ingots and sawing silicon wafers. The company chemically solubilizes its silicon nanocrystals (one-billionth of a meter wide) in ink, adds impurities (a process called "doping") to get the right electrical properties, and prints the ink onto any surface with an off-the-shelf industrial printer. While silicon normally melts at temperatures above 1,400°C (2,550°F), at 2 nanometers it melts at 300°C (570°F), cool enough to print onto stainless steel films. Though by the end the material looks almost like crystalline silicon, its ability to harvest light energy is vastly improved. At nanoscale, silicon can be made to perform like many semiconductors in one. By changing the size of the particles, called quantum dots, Innovalight can tune them to tap the sun's full spectrum, which conventional silicon only partially uses.

Right now Innovalight's prototypes look like old-fashioned rolls of Kodak film, but by the end of 2009 the company aims to produce each year enough flexible solar material to generate 100 megawatts at the unimaginably cheap price of 30 cents a watt. In the space of five months in 2007, it pushed efficiencies from 2 percent to more than 9 percent, meaning that a percent of incoming solar energy comes out as electricity.

Most remarkably, Innovalight's silicon quantum dots seem to have found the solar holy grail. Although many photons carry enough energy to unleash several electrons, photovoltaic materials have never been able to produce more than one excited electron for each incoming photon (instead, the excess energy is squandered as heat). But in a July 2007 paper in the American Chemical Society journal *Nano Letters*, Arthur Nozik and a team of scientists at the National Renewable Energy Laboratory confirmed that Innovalight's quantum dots are the first to get "multiple exciton generation" in silicon nanocrystals. While the theoretical maximum efficiency for a crystalline-silicon cell is 33 percent, Nozik calculates that this breakthrough could push Innovalight's efficiencies to

44 percent and as high as 68 percent with concentrated sunlight. As yet, the electrons have only been seen, not harvested; Innovalight still must figure out how to extract those extra electrons to generate electrical power. "But should this approach prove technically viable," concluded a 2007 Deutsche Bank report, "it would eclipse the conversion efficiency aspirations" of all other thin films.

Burke appears to have the fearlessness for such a gamble. His early career is a blur of leaps, each of them to bigger responsibilities and more cutting-edge experiments. He was twenty-one and studying physics at Trinity College, Dublin, when he saw a job posting to work with optical lasers and amplifiers at the Nippon Electronics Corporation (NEC). He took an immersion course in Japanese and by summer's end had moved to Tokyo. It was 1989, and Japan was an electronics powerhouse at its peak before the bubble burst—as he says, "a glorious time to be in the country surrounded by technology." He began publishing papers and making a name for himself, so much so that the Irish embassy arranged with Trinity to award him a masters degree for his NEC research. "It was a very economical way to get a masters, getting paid a salary and having my trips to Ireland financed. Very efficient."

He stayed with NEC for three years but found research and development too introverted. "I definitely did not fit the mold. I like interacting with people. I was twenty-two, single, enjoying all the attention, traveling to China, Korea, and Indonesia, a bit wild. It was great fun. One of the surreal things about being part of a small minority in Tokyo was that anytime anyone famous came from home—the prime minister or a rock star—the Irish ambassador would round us all up for the party."

In 1992 AT&T hired Burke away from NEC to do product marketing for its microelectronics group, then asked him to move to the United States. Six months after his arrival, big AT&T came to an end: the three-hundred-thousand-person company was dismantled and split into three independent companies, and Burke's division became Lucent. Within a few months he had been promoted to Lucent's director of marketing for Europe, the Middle East, and Africa, and moved his family to Germany ("my children have many different passports"). Three years later he returned to Allentown, Pennsylvania, to run the company's $400 million opto-electronics group. He was only thirty-two, the youngest director ever at AT&T and Lucent.

So he took another leap: in 1999 he quit his job to become senior vice president for marketing and business development at a small San Diego optical switching start-up called OMM. "I wanted to build a company and participate in the explosive growth of telecom, and they had a really cool new technology," he says. By 2001 the company had raised $135 million; filed to go public, with Credit Suisse as underwriters; and started its road show, with its product qualified in all networks, $40 million in orders, and the presidents of Selectron and Gateway on its board.

Then the market unraveled, and the orders and initial public offering disappeared. "We had to manage ourselves through that very hard downturn," Burke recalls. "We stayed intact, finally sold our IP. From a financial point of view it was not an excellent outcome.

"Silicon Valley forgives failure, if you get up and dust yourself off. In the end, I think it helped me gain credibility. Everyone can do well in an up market where everything's shooting toward the sky. But I learned more in that downward piece, from '01 to '03—letting go of a lot of people, bringing down a very expensive operation— than I'd learned in my whole career.

"Though I would not like to do it again."

Sevin Rosen Funds, the early-stage venture capital firm that had been OMM's biggest investor, asked Burke to become a partner. He spent a year there. "They paid me very well, and I looked at great technologies, but I found venture capital a bit boring," Burke recalls. "You meet smart people with great ideas. But you listen from afar, have no real participation—you're just getting fed the update. I missed being involved in getting those results."

Eventually, Sevin Rosen asked Burke to run Innovalight, an early-stage Minnesota-based company. The first thing he did was move Innovalight to California. "Capital is much easier to raise in California. And the Silicon Valley culture rewards risk-taking. People don't hesitate here taking chances with their career and jumping into unproven technology, knowing it may not work out. I've lived in the UK, Japan, Germany, the East Coast, and this twenty-mile radius is unlike anything I've ever seen. I'm having an absolute ball, and it's the smallest thing I've ever been involved in."

Burke has assembled a global team of two dozen, including fourteen PhD physicists and chemists from Italy, Belgium, Mexico, Russia, Ukraine, China, Taiwan, and Greece, who together write about two patents a month. Alf Bjørseth, a Norwegian who founded and until recently ran the world's most vertically integrated solar company, with a $20 billion market capitalization, recently joined Innovalight's board. Burke wants to grow the company to fifty people by 2008, and would like to tap the deep expertise concentrated in Germany and Japan, but is hamstrung by U.S. immigration policies. "The U.S. could get the best people in the world, superstar PhDs educated at the expense of taxpayers in other countries, but they're not allowed to stay." Because immigration is easier for Australians, he is hiring from the University of New South Wales, another powerhouse in solar energy. In October 2007, Innovalight announced that it had raised $28 million in new capital.

Growing concerns about the safety of nanotechnology may present hurdles to Burke and others in the field. Since 2003, Environmental Defense Fund has been involved in a worldwide effort to strengthen the oversight of nanomaterials. In general, regulators treat these materials as if they were identical to the same chemicals in bulk form, despite the fact that the nanoscale versions are valuable precisely because they behave in radically new ways. The limited data now available suggest that some nanomaterials may be mobile and long-lasting in the environment and organisms, and may be capable of damaging brain, lung, and skin tissue.★

For his part, Conrad Burke seems surprised by questions about safety, which is perhaps understandable given his neighborhood's general faith in new technology. He says his workers are thoroughly protected, and that Innovalight's nanoparticles are essentially erased by the time the solar thin film is complete. "We use nanotechnology as a cheap vehicle for manufacturing, but by the end the nanoparticles are stuck together, deformed in processing in a way that they're not ever able to be released."

BURKE'S BIGGEST CHALLENGE, as for most innovators, will be taking his team's success in the lab out onto the production floor. One of his favorite people—and a model for the challenges he knows lie ahead—is Dave Pearce, the founding (and now former) CEO of thin-film competitor Miasolé, which in early 2007 seemed on the brink of commercial production at its 111,000-square-foot Santa Clara factory and subassembly facility in Shanghai. (Pearce, who is paunchy, balding, and middle-aged, gave Miasolé its quasi-Italian name—a loose translation is "my sun"—soon after the film Under the Tuscan Sun was released, in a vain attempt to lure actress Diane Lane onto his board. "We did get John Doerr," he jokes.)

Miasolé does not use silicon but a compound semiconductor

★ In an unusual partnership, DuPont and Environmental Defense Fund have developed protocols for assessing the risks of developing, producing, using, and disposing of nanomaterials. The goal is to reap the full promise of this technology without unintended consequences (think asbestos) or a public backlash (think genetically modified organisms). DuPont has made the framework part of its mandatory product stewardship process.

made up of copper, indium, gallium, and selenium known as CIGS, which scientists have long tried to exploit. A CIGS film as thin as 1 micron has the same photovoltaic effect as a typical crystalline-silicon wafer 200 to 300 microns thick (about the thickness of a human hair), translating to savings both on expensive semiconductors and on weight. CIGS also works better than silicon at low angles of the sun and on hazy or cloudy days. And like Innovalight's thin film, it can be churned out rolls at a time.

Unlike the youngsters running many of the solar start-ups, Miasolé's founding team brought decades of experience at industrial-scale manufacturing. Pearce himself arrived in Silicon Valley in 1985, when he became CEO of Domain Technology, a hard-disk maker that had burned through $22 million of its $23 million and was about to go under. In six months Pearce turned the company around, proving out its "sputtering technology" for depositing magnetic films and turning a profit. (Portions of Domain were later acquired by Seagate.)

Pearce likens the sputtering process, which Miasolé now uses to make solar films, to a game of billiards. Inside a vacuum chamber, magnets are used to accelerate argon ions (the cue ball), which then knock off atoms (the pool balls) of the target material (copper or indium on these solar films). The atoms settle one by one, producing precise films as thin as five to ten atoms.

In Miasolé's sparkling Santa Clara factory, that process takes place in giant U-shaped machines. At one end, a meter-wide roll of flexible stainless steel unscrolls at the rate of two feet a minute; at the other end, out comes the photovoltaic foil. Each production run delivers several square miles of solar foil, which is then encapsulated in a rugged, flexible material. The company builds its own capital equipment to save money (they spend a tenth as much as if they bought it, says Pearce) and to make possible continuous redesign for ever-higher throughput.

Pearce grows particularly animated when he talks about production challenges—for instance, how to get the grid lines, which conduct the electricity, onto the cells. When engineers tried printing silver ink directly onto the film, the ink wicked down into microscopic divots in the foil and shorted out the cell. They realized they needed to bridge the divots, so they designed a decal, with the gridlines printed onto a polymer, which they then stuck onto the cell. The decal, in turn, gave birth to other inventions. By printing the decal with black ink first, they eliminated visible grid lines, and reduced reflection so it absorbs more solar radiation. They also found they could hang the buss bar, which transports the electricity, from the decal's edge, leaving the whole cell photoactive.

Miasolé's aim is to make solar technology "simple enough for your average Home Depot customer to do it themselves," as Pearce put it. Their panels will include mounted sensors and wireless communications capability, to make them smart enough to tell their owners when they need to be cleared of fallen leaves. Pearce explains: "In most systems, if shade falls on one cell you degrade function in the whole series. Installers struggle to avoid that with highly precise positioning. But because these will have discrete electronics for each, a problem with one won't mess up the rest."

At the beginning of 2007, Pearce was certain that by 2008 Miasolé would be producing 200 megawatts of new generating capacity each year, at a cost of $2 per watt installed. That "installed" part is important. Unlike fragile crystalline-silicon cells, these photovoltaics needn't be framed in glass and aluminum and mounted on the roof: they *are* the roof. The product bypasses the costs of framing and installation, and also the conventional solar distribution channels, which have become badly bottlenecked. Instead, Miasolé's panels will be part of the building-materials industry, like sheetrock and tarpaper. Encapsulated into composition shingles, which already make up 85 percent of U.S. roofs, they can be glued right on to plywood roof decks all over America. The company has already begun discussions with large-scale commercial and residential developers like Toll Brothers and KB Home, which could offer an electricity-generating roof bundled into the mortgage, like a granite countertop or high-speed Internet access. Because the owner will need much less power from the grid, Pearce says, his combined mortgage and electricity bill will be less with the solar roof than without.

As a member of SolarTech, a new consortium that includes SunPower (a leading crystalline-silicon solar cell maker) and PG&E and aims to turn "Silicon Valley into Solar Valley," Pearce has worked to expand such financing options to eliminate a chief obstacle to broad consumer adoption. People buying rooftop solar systems today are essentially required to prepay a couple of decades' worth of energy bills; Pearce compares it to buying a car and having to pay upfront for many years' worth of gasoline.

To fix that, various financial institutions have adapted "power purchase agreements," a common financing instrument for conventional electricity generation, to solar power. Morgan Stanley, for instance, is financing the installation of SunPower solar arrays at twenty-two Wal-Mart facilities. Wal-Mart makes no capital investment and does not own the solar panels. Instead, it has committed to buy the electricity generated by the panels at favorable rates, which are locked in for twenty years.

Bank of America has a similar arrangement with Chevron Energy Solutions and the San Jose public schools. The bank finances and owns the solar installation, earning state and federal subsidies and tax credits. Chevron installs and operate the modules, and the school district buys the green energy at below-market rates, reducing its demand for utility-supplied electricity 25 percent and saving $25 million over the life of the equipment. A few companies offer the full range of services. SunEdison, for instance, finances, builds, owns, and operates rooftop solar installations for Whole Foods and Staples. Industry insiders call it a "distributed utility."

To make such arrangements available to smaller businesses, SolarTech is creating an online marketplace to match consumers with financing and solar suppliers. Eventually, the group envisions individual homeowners being able to switch to solar without any front-end cost, cutting both their global warming impact and their electricity bills.

The technologies themselves are proving more stubborn. In May 2007 Pearce announced delays in commercial production. "We're trying to give birth to a new process. The trouble is that we don't know how long the gestation period is." Though Miasolé's research and development lines were still hitting the target efficiency of 8 to 10 percent, that was only on 5-square-foot panels of film; on its big commercial lines, the company was getting just 4 to 6 percent efficiencies—a rate that improved somewhat by October 2007. Miasolé is also contending with limits on the global indium supply, which *PHOTON International* in July 2006 estimated as enough for just 4 gigawatts of CIGS. That scarcity has already affected the semiconductor's price, which in 2007 was triple what it had been five years earlier.

In September 2007 Pearce became chairman of the Miasolé board, yielding the job of chief executive to Joseph Laia; by year's end, Pearce and most of his executive team were gone. Laia was most recently a group vice president at KLA-Tencor, which makes diagnostic tools to increase semiconductor manufacturers' yields; in 2007 the company had revenues of $2.7 billion.

While confronting the challenges of going to scale, Miasolé and Innovalight are well aware of rivals like Nanosolar, which in late 2007 shipped its first CIGS thin films. Asked about silicon quantum-dot start-up Stion, which raised $21 million in its first year, and Octillion, which says it can spray a transparent coat of nanosilicon film onto glass to make windows that can generate electricity, Conrad Burke shrugs. "They just started working on it, and they're claiming they went from zero to sixty in ten seconds? We've been on it for four years. We know all the processes that don't work because we've tried them. Silicon is a beast. And before you can build the technology you have to build the tools you need. It's not like designing an iPhone, where all the pieces are available and you just have to put them together. This stuff is hard and takes longer than you think. People think Google was an overnight success, but it took ten years."

And while it is tempting to judge the likelihood of these newcomers' success by how much money they have raised from venture capitalists, Burke warns against it. "The VCs have more money than they know what to do with," he says. "But for every dollar you raise, you need to return a multiple of that. If you raise $100 million, you have to return a billion to your venture investors. That makes it a little more difficult."

One thin-film company has already had a fairy-tale ending, which is often invoked by other companies still struggling to get the process right. When Phoenix-based First Solar was founded in 1999 with $250 million from Wal-Mart family member John Walton, who was passionately concerned about global warming, it planned to spend three years and $40 million to commercialize its process for condensing cadmium telluride gas onto glass. Six years and $100 million later, the company was still trying. Meanwhile, BP Solar abandoned its own large-scale effort to commercialize cadmium telluride. Concerns about cadmium's toxicity and potential release in a fire or landfills further delayed First Solar's development and added the ongoing cost of a company-run program for recycling panels.

By 2006, nonetheless, First Solar had secured the prize. Revenues that year reached $135 million, up from $13.5 million in 2004. Its initial public stock offering raised $400 million; shares issued at $20 were worth $74 by the following June, bringing the company's total value to $5 billion. By 2007 the company's plants in Ohio, Germany, and Malaysia could make a two-by-four-foot glass module in two and a half hours, with efficiency exceeding 9 percent. (Though their modules sold in 2007 for more than $2 a watt, they expect to cut costs enough to get the market price down to $1.) Annual revenues were forecast to more than triple, to $480 million, and the stock was trading well above $200. The company had long-term contracts worth $1.62 billion to supply modules capable of generating 795 megawatts to European and Canadian buyers—nearly eight times the total shipped in 2006 from every solar factory in the United States—and was planning to raise as much as $1 billion in another public stock offering.

BILL GROSS, CEO AT ENERGY INNOVATIONS, is as interested as the thin-film makers in generating trillions: both the trillions of watts of clean energy the world needs, and the trillions of dollars to be made in the energy markets. "Energy drives 50 to 75 percent of our economy. It's embedded in everything." He picks up a water bottle. "It's embedded in this. And it's a complete commodity market. All electrons coming out of the wall look the same."

But rather than make acres and acres of cheap photovoltaic material as Innovalight and Miasolé are doing, Gross believes the way to get to the terawatt scale is by concentrating the sun. "The sun is just too diffuse," he explains. "It radiates 1,000 watts per square meter of the earth's surface. A hair dryer is 1,000 watts per square inch. So the sun is sixteen hundred times more diffuse than a hair dryer." To make 1 gigawatt of power (enough to power San Francisco) using traditional photovoltaics requires four square miles of silicon, according to the National Renewable Energy Laboratory. Concentration reduces that area by orders of magnitude: under Energy Innovations' sun concentrators, Gross says, 1 square inch of photovoltaics will produce as much energy as 800 square inches without concentration.

Concentration is also, he believes, the cheapest of all options because it leverages the scarce supply of purified materials: "If you concentrate the sun a thousand times, a gigawatt's worth of silicon will get you a terawatt." Optical devices for concentrating light, and trackers to follow the sun across the sky, are cheaper than the photovoltaics themselves. "Thanks to Moore's law," says Gross, "a full-fledged microprocessor that can track the sun costs 20 cents, down from $2,000 just twenty years ago."

Gross has been toying with concentrating the power of the sun ever since high school. In 1973, at age fifteen, he began making small parabolic dishes in shop class to power Stirling engines, which use heat to drive a piston. He sold the devices through mail order ads in the back of *Popular Science* magazine. "Super solar devices," the ads read. "Catalogue 25 cents." In 1996, Gross founded Idealab, a Pasadena-based incubator for high-tech start-ups. In a decade it has spun off forty companies, including Internet directories (City Search) and paid search sites (Goto.com). In 2003, Idealab sold Overture Services to Yahoo for $1.7 billion.

Where Dave Pearce would fit right in at a midwestern factory, it is hard to imagine Gross anywhere but California. He has a dreamy, slightly otherworldly air, like a boy prodigy immersed in his own imaginings. He talks quickly, sometimes drawing pictures to explain himself (his paintings hang on Idealab's white brick walls). All of Idealab's incubating companies occupy an open workspace that is a study in high-tech cool, with exposed metal conduit and a wide-open "village" of bright yellow workstations meant to encourage creative mingling. In the reception area, an "Energy cam" tracks the performance of the solar modules on the roof, including the kilowatt-hours generated and the pounds of carbon dioxide avoided since December 2006. A quote from Schopenhauer flanks the entry:

"All truth passes through three stages: First, it is ridiculed. Second, it is violently opposed. Third, it is accepted as self-evident."

Energy Innovations' initial concentrator design, the Sunflower, arrayed a circle of mirrors (called heliostats, because they track the sun) around a Stirling engine. That design was shelved in favor of a lens that could produce higher concentrations by focusing sunlight down through a glass funnel and onto a solar cell. The heliostat technology was taken over by another Idealab company called eSolar, which couples the mirrors with small towers and thermal receivers to make solar power plants, the subject of Chapter 3. In late November 2007, Google named eSolar a partner in its initiative to spend hundreds of millions of dollars developing "renewable energy cheaper than coal" (the program's name uses mathematical shorthand: RE<C).

The new Energy Innovations modules, which resemble banks of stadium lights laid flat on the roof, come fully assembled for easy installation. Twelve are linked on a frame that pivots up and down and across the horizon; microprocessors guide the movement, keeping the lenses and cells aimed directly at the sun. Energy Innovations designed both the tracking software and much of the hardware—from the chemistry of the plastic and patterns of grooves within the lens, to the gill-like passive cooling system, designed to conduct away the intense heat while using minimal aluminum. In an environmental test chamber, the modules are hammered with one-inch balls of hail and hundred-mile-an-hour winds, to make sure they do not lift off or break or get wet and short-circuit. The commercial modules, which will be made in China, will "origami down" into a compact form to reduce shipping costs.

The real magic comes from coupling the concentrators with the world's highest-efficiency solar cells. The cells don't come from Silicon Valley, but from a big, old company, Boeing subsidiary Spectrolab, which for two decades has made the photovoltaics that power NASA satellites and lunar explorers. Spectrolab is now bringing that space technology back to Earth, developing "terrestrial applications" for its cells. In December 2006 it set a new world record of 40.7 percent efficiency, the highest ever achieved by any kind of solar cell.

In a big old building in a run–down industrial area east of Los Angeles, Spectrolab technicians in clean rooms feed elements to a crystal growing atop a wafer, layering gallium, indium, and arsenide, each of which absorbs a different wavelength of light. Each layer is extremely thin and semitransparent, allowing light to pass through to one of three active junctions, where the different wavelengths are absorbed. In darkened labs, researchers test the spectral performance of new compounds, trying to push efficiencies higher still. Others work on removing the wafer, to make the cells lighter and flexible, like a thin film.

With its $50 million infrastructure and decades of experience in space, Spectrolab is one of the few places capable of making such super-cells. The difficulty is that the elements that work best together to maximize the conversion of solar energy into electricity do not easily grow together because their atoms are differently spaced. "It's like trying to put a big peg in a small hole," explains Nasser Karam, Spectrolab's vice president for advanced technology products. "It sits uncomfortably." To ease the strain, Spectrolab adds twenty-seven buffer layers between the sheets of semiconducting alloys, which allow the atom spacing to change slowly, like a series of frames in cartoon animation.

In a big assembly bay, immense barrel vaults and the flat broad wings of satellites wait for their cells to be affixed. On the wall hangs a gift from NASA's Jet Propulsion Lab: pictures of the Mars Rover chugging along, powered by Spectrolab cells on its flanks—cells that not only have withstood temperature changes from 100°C to

−170°C (210°F to −340°F) in a few seconds but also, to everyone's surprise, have been undeterred by Martian dust.

By the time it is finished, each Spectrolab wafer can make a kilowatt of power. Every two days, the facility makes one thousand wafers—a megawatt of generating capacity. By late 2007 the company had orders in house for a million wafers—a gigawatt. Among its customers is Concentrating Technologies (C-TEK) in Alabama, which is building a grid-connected "solar farm" for Arizona Public Service.

Most companies combining concentrators with photovoltaics are pursuing the utility-scale market, but Gross prefers to compete with retail price. He is focused on the commercial market, half a megawatt or more, to leverage the expense of installation. "To power your building you typically need a watt per square foot, and at 15 percent efficiency, as much roof as floor. For instance, at Idealab we have forty-five thousand square feet, so we need 45 kilowatts, and we can get that if we cover our whole roof at those efficiencies. If your roof is pitched, you can only cover the south half. But we think there are enough big, flat roofs out there for us to make a big impact on peak demand. You can't turn off the coal plants but you can make it possible to build no new ones."

To break through the sales and distribution logjam that hampers many solar start-ups, Energy Innovations purchased a big power and systems integrator, now called EI Solutions. In 2007 the company installed enough Sharp crystalline-silicon cells on three acres of roof at the Googleplex in Mountain View, California, to produce 1.6 megawatts, at the time the biggest commercial installation in the United States (Google cofounder Larry Page is an investor in Energy Innovations). Google expects to save almost $400,000 annually in energy costs, paying for the system in seven years. Real-time information about the system's output is posted on Google's corporate Web site. For Puget Sound Energy in eastern Washington, EI Solutions arrayed solar panels capable of generating half a megawatt between windmills, getting the land to do double duty. The beta installations—the first real-world test—for Energy Innovations' own rooftop concentrators will be in Pasadena.

Meanwhile, in laboratories around the country, photovoltaic innovation continues at a fierce pace. Seven months after Spectrolab set its record, for instance, a Defense Advanced Research Projects Agency and University of Delaware consortium that includes BP Solar, EMCORE, SAIC, National Renewable Energy Lab, and MIT achieved 42.8 percent solar-cell efficiency. Instead of layering different semiconductors onto a single cell, researchers used "spectral splitting" optics to divide the sunlight into three "energy bins" of like colors, directing each onto a different type of cell capable of absorbing that range of wavelengths. The device's wide acceptance angle, they claim, eliminates the need for tracking devices. DuPont will help transition the lab-scale work to a manufacturing prototype.

Yet another strategy is under development at StarSolar, a Cambridge-based company that was launched by a young MIT graduate student named Peter Bermel when he won a $100,000 prize for his business plan. Bermel uses "photonic crystals," which can be tuned to reflect specific wavelengths to recycle the photons that slip through silicon, making possible much thinner wafers. Backing the cell, the nanoscale structures also bend the light so that it reenters at an oblique angle and bounces around inside, upping the odds that the silicon will harvest its energy.

Caltech chemistry professor Nate Lewis is looking beyond thin film to solar materials that would scale up "like rolling out carpet or painting your house." His solution to the thickness problem is a new geometry. What if a solar cell were tall but very thin, like a nanoscale blade of grass or Berber carpet fibers? Then it would be thick enough from top to bottom to absorb the solar radiation, but thin enough so that the excited electron could go out sideways. Lewis says that "it opens up the possibility of using cheap materials that could never be purified enough to get the electrons to the top." A variant might use what Lewis calls "solar sand"—nanoscale balls of titanium dioxide, a cheap material used in toothpaste and white paint; they would be suspended in a liquid conductor that could then be painted onto a house.

THESE PHOTOVOLTAIC INVENTORS and entrepreneurs all know what happened in the 1980s when oil prices dropped after their 1970s surge and both public and private commitment to developing solar energy pretty much dried up. This time, they think the chances for technological and environmental breakthroughs are far greater. In the words of Conrad Burke of Innovalight, "Twenty years ago no one would have put money into these things. They would have viewed it as some niche-y environmental thing. Now it's for real."

At the same time, all these entrepreneurs agree that serious changes in government policy are needed. Until now, the main policy tools used to advance alternative energy have been subsidies and mandates, which have helped promote renewable energy. Although Japan and Germany account for just 10 percent of the global energy market, thanks to long-term subsidies they command 70 percent of the solar market worldwide. Germany has used a "feed-in tariff," which requires utilities to buy electricity from renewable-energy producers, including owners of rooftop systems, at above-market rates. In 2007 German utilities paid up to 57 euro cents (about 72 cents American) per kilowatt-hour for solar energy, about triple the price at which they sold energy back to consumers. Not surprisingly, that has spurred huge investments in solar products. "All the high-value solar companies in the world have made it by selling their products in Germany," Burke says. "That country will be so wired up to the sun many nations will be envious. It's like Korea, which decided to wire the country with high-speed fiber-optics and now has the fastest communications network in the world—it makes ours look like bicycles."

But those successes are a testament to the long-term commitment Japan and Germany made to alternative energy—rather than to the particular way they chose to promote it. Subsidies and mandates have several critical weaknesses. For one, they depend on a degree of detailed knowledge and a prescience about technology beyond the reach of government regulators.

They also reward lobbying prowess more than technologies that actually perform, and can result in perverse outcomes. "The European model of feed-in tariffs richly rewards certain players," says John O'Donnell, a solar power entrepreneur you will meet in the next chapter. "And it creates bizarre situations which have nothing to do with slowing climate change." He points to the German subsidy for photovoltaic power as a prime example. Because Germany is a generally unsunny place, it takes as much as six years for a photovoltaic cell to generate as much electricity as it took to manufacture it. Demand for fossil electricity, therefore, has yet to drop, says O'Donnell: "Not a single coal plant has yet been shut down by this initiative," even while the net cost of electric power (including the big government subsidies) has risen to 50 cents per kilowatt-hour.

Subsidies are also notoriously fickle; they can get the ball rolling but also drop the ball when they are abruptly eliminated, as they so often have been for renewable energy. O'Donnell's colleague Peter Le Lièvre points to Spain, which pays a premium of about 25 cents American per kilowatt-hour for solar power. With its ample sun, "Spain is a delightful hotpot right now," says

Le Lièvre. "But it is an aberration of a single policy setting that could be turned off at a moment's notice." In the U.S., "the federal investment tax credit is very nice, but is ultimately a destructive weapon in that it's not a permanent market reform, but a bit-by-bit handout, as if we're begging for assistance. Nothing could be further from the truth. Market reform is a much more durable and sustainable platform on which to build our long-term investments. We strongly believe that mobilizing capital markets is the best method for deploying these technologies rapidly."

Policy, in other words, must harness the most basic and visceral impulse of capitalism: the pursuit of profit. Says Burke: "We'll start taking out coal and oil, and the end result will be good. That makes us more motivated than if we were just designing some new process for a laptop to make the Web go faster. But my primary goal is to return a lot of money to my shareholders, including me and the employees. The scaling of solar is not going to happen with a bunch of environmentalists. You need big, serious money." As *Economist* writer Vijay Vaitheeswaran puts it, "You link the technology to the capital, and that's where the rubber hits the road."[*] Events in California since its passage of a law capping carbon emissions offer proof of that, suggests venture capitalist John Doerr: "The biggest impact has been the availability of capital. Some would call it a bubble—I would call it a boom."

Bill Gross of Energy Innovations predicts that since "there will be ninety-nine failures for every one success," it will not be large, established companies that achieve the "disruptive innovations" that are necessary: "It's a question of small and nimble versus big and powerful."

What is the best way to encourage the small, nimble risk-takers? Gross argues for using tax policy to spur change. Fossil energy, he

---

[*] In an April 2007 interview on the PBS television program *Nova*.

says, should be brought to its true price with carbon taxes and similar levies—"then let the market figure it out." But nowhere in the world has a tax successfully solved an air pollution problem. And given the enthusiasm with which politicians generally greet taxes, it is unlikely to be the right solution to this one. More fundamentally, a tax fails to provide the certain pollution reductions offered by a cap: it tells polluters how much they must pay to pollute, but does not impose a legal limit.

That brings us to the cap-and-trade system—the best way to harness market forces to fix a market failure. Instead of forcing polluters to pay certain prices or to back particular technologies, the cap-and-trade system mandates only the pollution limit, then lets the competitive machinery of the market figure out the cheapest, most efficient way to get there. Mobilizing the market ensures that the hunt for the cheapest technologies will be as broad as possible, ranging as far as the human imagination; only with such a far-reaching search will the United States be able to reach the 80 percent reduction in global warming emissions that scientists tell us is necessary to stabilize the climate. That broad hunt, in turn, sets in motion a valuable cascading effect: as the market finds the most efficient technologies, and quickly brings down the cost of reducing pollution, the political will builds for even steeper carbon cuts—without the backlash that inevitably follows when the government tries to pick technologies and too often makes the wrong choice.

Dave Pearce, founder of Miasolé, is one of the growing number of clean-energy innovators arguing for this view. As one of fifteen members of the TechNet Green Technologies Task Force, he has been advocating in Washington for a federal cap on carbon emissions to "create an environment in which new energy technologies can emerge and thrive."

A task force document spells out its top priorities: "Climate change policies must internalize the environmental cost of emitting greenhouse gases, thereby reducing or eliminating the price differential between high carbon-emitting activities and more environmentally sound activities. A cap-and-trade system built around market mechanisms will be the most powerful driver of demand for new technologies." That dollar-a-watt threshold for solar energy to compete with fossil-derived energy, in other words, assumes that coal plants are not required to clean up their global warming pollution. Put a cap on carbon, and it has a price for everyone, including coal burners—and solar power becomes competitive in much less time. John Doerr says that if a carbon cap were enacted in 2008, Miasolé's thin-film technology would achieve "grid parity"—make electricity for the same average price charged by utilities—by 2010.

The goal, the urgent necessity, is to reduce global warming pollution in the atmosphere enough to pull us back from the precipice before the changes in earth's ecosystems and weather patterns become so rapid and so vast that we will no longer be able to reverse the catastrophe.

CHAPTER 3

Harnessing the Sun, Part II

In 2002 a soft-spoken physics professor named David Mills and a straight-talking industrial designer turned venture capitalist named Peter Le Lièvre launched a company in Sydney, Australia. Their aim was to drive steam turbines with sunshine—to build a power plant fueled by the sun that would be as big and cheap as those fired by coal. As inventors often do, they rented a suburban garage, gathered money from family and friends, enlisted a student Mills knew, an engineer named Andrew Tenner, to work for free, and set about assembling a mirror and steam pipes to generate electricity.

They miscalculated on one front: the finished contraption was as big as a semitrailer, and they almost didn't get it out of the garage. But after they managed to haul it out into the sunshine, it worked. "It would burn the hairs off the back of your hand," says Le Lièvre. By 2004 they had scaled it up enough to make their first megawatt. "In three hours, we verified three years of work," Le Lièvre recalls. "It was light and strong and beautiful to look at, and that first megawatt set everything aflame."

For Mills, that moment culminated almost thirty years of work. Born in Canada, he had moved to Australia and earned a doctorate studying how prisms could curl light onto surfaces designed to convert it into usable heat or electricity. By the 1990s he was running the solar thermal research lab at the University of Sydney. Among the earliest inventions he helped develop was a simple solar hot-water heater using vacuum tubes (like thermoses). It is now in 30 million Chinese homes. (His partner in that early enterprise, Huang Ming, became chairman of Himin Solar Energy Group, which in 2007 had seven thousand employees and was growing by nearly 40 percent a year. Dezhou City, where its 200,000-square-meter facility is located—it looks like a jumbo-jet factory, says Mills—has been named China's Solar City.)

At about the time Mills was inventing his solar heater, an Israeli-American company called Luz was beginning construction on a new kind of solar power plant in California's Mojave Desert. Over several square miles, Luz engineers erected long, curved mirrors called "parabolic troughs," which track the sun as it makes its daily journey across the sky. The mirrors focus the sun's heat on double-layer pipes—specially coated stainless steel enveloped by glass, with a vacuum between the layers. These "receivers," which are filled with synthetic oil, run horizontally within the arc of the troughs and, attached to the mirrors, go along for their daily ride. The concentrated sun heats the oil, which is then pumped to a boiler to make steam. From there, it is like any big fossil-fuel plant: the steam drives a turbine and generates electricity. To maintain the steady high temperatures required by their turbines, the Luz engineers added a natural gas burner, which keeps the steam at 400°C (750°F) even in the soft light of morning or under gathering clouds, but supplies less than 2 percent of the power. The system runs only during daylight hours.

If photovoltaics are the offspring of the Information Age—riding on the shoulders of advances in quantum physics and solid-state electronics—solar thermal electricity remains rooted in the Industrial Age: big and centralized, built of concrete and steel, with lots of moving parts and belching boilers and cranes and mechanical engineers. Yet it is the only clean-energy technology with the near-term potential to match coal not only in price (which some windmills can do) but also in scale, which is why—after nearly two decades of dormancy—solar thermal start-ups are scoring major, long-term contracts with utilities, gaining access to core transmission lines and beginning to secure giant amounts of money— hundreds of millions of dollars.

Between 1984 and 1990, Luz built nine solar thermal plants in California's Mojave Desert with a total capacity of 354 megawatts, enough to power 350,000 homes. From a distance, they look like an immense shimmering farm in the desert planted with seemingly endless rows of some silvery crop. The plants are still operating today. In two decades, they have supplied more than 11,000 gigawatt-hours to Southern California Edison and have generated almost $2 billion in revenue. By 1991, however, Luz was in bankruptcy. According to the National Renewable Energy Laboratory, changes in the federal policies that had provided a favorable market environment in the early 1980s contributed to its collapse: "Beginning in 1986, the 10 percent energy tax credit for solar energy property was extended in a piecemeal fashion, anywhere from nine months to two years at a time, creating tremendous financing uncertainty." In 1990 the company had to build its ninth plant in seven months to qualify for the credit, leading to cost overruns that exceeded revenues.

As Luz was selling its assets, Mills was experimenting with ways to simplify trough technology to cut costs. Instead of curved glass, which is expensive to cast, he designed a "Fresnel reflector" from long strips of flat glass, arrayed side by side, which cost a third as much and can be manufactured out in the desert where they will be used. (The idea of making a curved mirror or lens by combining flat facets of glass was invented in 1822 by a French physicist named Augustin-Jean Fresnel, and has been used in lighthouses ever since.) Mills placed the mirrors nearer the ground, to reduce the amount of structural steel needed; in storms, they turn their steel backs to the wind and hail. He suspended the pipes farther up in the air, on their own fixed supports. This gave the system a longer focal length, which in turn allowed him to relax the reflectors' curve; the separation of pipe and mirror also allowed him to give each reflector the choice of two receivers, reducing shadowing and allowing a smaller footprint for the solar field. Instead of oil, he used water in the pipes, running the boiler pipe right through the receiver to make steam directly with the sun, without heat exchangers. "Innovation is not one thing," says Mills, "but many, many small things." After fifteen years of such engineering, he had brought the cost per square meter of a solar field to just over $100, less than half the cost of parabolic troughs.

Mills made trade-offs to achieve those savings: the steam produced by his "compact linear Fresnel reflector" is cooler, just 200°C to 300°C (390°F to 570°F), and wetter; the technology's solar-to-electric efficiency is just 10 to 15 percent. The newest parabolic troughs turn about 14 percent of incoming solar energy into electricity; one competitor, BrightSource Energy, says its first-generation power-tower technology will achieve 20 percent efficiency and the next generation, using higher temperatures, will reach 24 percent.

But lower temperatures also brought additional cost savings to Ausra, which is what Mills's company is now named. (Ausra is an ancient Indo-European name for the goddess of the sunrise; the word survives in modern Lithuanian and means dawn.) Instead of the "supercritical" turbines developed for the coal and gas industries, which use high heat and pressures to reach high efficiencies, Ausra will use saturated steam turbine technology from the nuclear power industry, which, Le Lièvre explains, "has spent many years and huge sums developing turbines to run on low-temperature wet steam, with no superheating stage." Because nuclear turbines can handle varying steam pressures, Ausra will not need expensive control systems. And whereas Luz had to superheat steam with natural gas to keep a steady temperature for its turbines, Ausra will not: its system can tolerate fluctuations, as the sun comes and goes on cloudy days. That saves on the gas burner and the gas, as well as on thick-walled tubing, specialized heat transfer fluids, and ultradurable coatings, none of which are needed.

A KEY ADVANTAGE of solar thermal over photovoltaics is the capacity to store energy as heat, which is far cheaper than storing electricity. (Consider, for instance, the low cost of a lunch-box thermos—about $7—compared to that of a laptop battery—about $150. Both store about the same amount of energy, assuming the thermos is filled with boiling water. But while the filled thermos is mostly useful for keeping that boiling water hot, the laptop battery could be used to make light, turn a motor, or even to boil water.) Again, Ausra will opt for a lower-efficiency, lower-cost solution than its competitors: where most solar thermal electric plants use molten salts as their heat storage medium, which can go to 1,000°C, (1,830°F), Ausra simply condenses steam into hot water—lots of it. To get the lowest-cost energy, in fact, the company will build its solar field three times larger than its turbines' peak input, so that on summer afternoons two-thirds of the energy will go "into the tank," allowing Ausra to run generators more hours of the day and more days of the year. (Ausra says that it will incur capital costs of just $15 for each kilowatt-hour it stores, but has not yet proved that the system works at a commercial scale.)

How Ausra will actually store so much pressurized water is now a central focus of research. The most likely solution is to place metal-lined tanks underground, using the earth's insulating properties and the weight of the rock overburden to maintain the pressure; the company is studying designs that would bury the tanks anywhere from just below the surface to as deep as one hundred meters, keeping them away from aquifer layers, which would cause them to lose heat. Steam would then be flashed directly from the tank into the turbine. (Ausra could even "host" other renewables, like wind, which is often abundant at night when there's no demand. Converted to heat, wind energy could be stored overnight in company tanks and dispatched on demand.) For a 150-megawatt solar plant, the volume needed would be like a grain silo, ten meters in diameter and thirty meters tall. According to Mills's models, the price of electricity from a 700-megawatt plant with twenty hours of storage would fall to 7 cents per kilowatt-hour, which beats the cost of electricity produced by trough technology and is competitive with that generated by natural gas.

Ausra had been running a pilot project in Australia for a year, producing enough supplemental steam to make 12 megawatts at a coal-fired plant owned by Macquarie Generation, when John O'Donnell, a serial entrepreneur from Silicon Valley, and venture capitalist Vinod Khosla came knocking. O'Donnell had heard Steven Chu, winner of the Nobel Prize in physics, speak about global warming; that talk had launched him on a quest "to find a technology that could displace a billion tons of carbon by mid-century." O'Donnell and Khosla expressed their interest in Ausra's project. "I didn't know who Vinod Khosla was, so I was completely fearless," says Peter Le Lièvre. "I said something absurd like they'll have to offer us $50 million, thinking they never would—and then they did." A second round in 2007 brought another $25 million of equity financing to the company. As Le Lièvre explains, "A number of people in the U.S. had been after us to license our technology to them, but we wanted to be masters of our own destiny. John said, 'Well, let's set up the company [in the United States] and you still run it.' We said we thought the U.S. was all George W. Bush and foot-dragging, and John said, 'No, in California and other states there's rapidly expanding interest.' "

In March 2007 the team moved to Palo Alto, families and all. They began rapidly hiring from the power sector, aiming to grow the company from thirty-five to seventy people. They opened offices in the sunny states, started lining up projects with California utilities, and initiated a search for a seasoned power executive to become their new CEO. Le Lièvre wanted to get back to working on the technology. "And the amounts of money we're deploying are so big, and the industry so established it makes sense to get someone from that environment," he explained. "Executives at the big power companies are tired of being tarnished as wreckers of the climate, and see in us a way to make good." By September 2007 the search was over: Ausra announced that Robert Fishman would become its new CEO. As executive vice president of Calpine, a power company in San Jose, California, that makes 3 percent of all the electricity consumed in the United States, Fishman had managed natural gas and geothermal power plants producing more than 24,000 megawatts.

The move from Sydney has been stressful, says Le Lièvre. "Those of us from Australia are kind of the-glass-is-half-empty types. We intentionally delayed launching a Web site and kept a low public profile so that we would have the luxury to fail in private. But John [O'Donnell] is a true believer, an evangelical who goes after utilities and politicians and regulators. The political maneuverings here are much more aggressive. And Vinod [who now sits on the Ausra board] is well known for driving people hard. He raised nearly $50 million in our A round and has had his foot to the floor ever since.

I wouldn't miss it for the world. At the end of the day, he has a great intuitive sense of market timing and deal making, takes a lot of risks and has opened a lot of doors. Love him or hate him, we will be better for his involvement—provided we can survive the next two years."

THOUGH SOLAR THERMAL TECHNOLOGY had languished for sixteen years in the United States, by the time of Ausra's arrival, competition was heating up. In 2006 a North Carolina company called Solargenix Energy had built a 1-megawatt parabolic-trough plant north of Tucson, Arizona, lining up six rows of mirrors, each nearly a quarter-mile long. In 2007, with Spanish partner Acciona, Solargenix secured $266 million in long-term financing to build Nevada Solar One, a 64-megawatt plant on three hundred acres in the Eldorado Valley. Nearly all the plant's components were made in Germany. Schott supplied nineteen thousand vacuum-tube receivers made of steel and a new type of glass with the same thermal coefficient as the steel, so the two materials react in unison to temperatures fluctuating from 400°C (750°F) to below freezing. Flabeg, also German, supplied the mirrored troughs; Siemens built the turbine.

"We're definitely in competition with our parabolic-trough cousins," says Le Lièvre. "Those are the companies we're catching up with, particularly the large Spanish companies, who can legitimately argue that they have the established and proven technology and are therefore getting more traction with project finance and with customers buying their plants. But their problem is price." Susan D. Nickey, chief financial officer for Acciona Energy North America, acknowledges that as a "first mover" in the solar thermal field, Acciona did incur higher costs, which had to be factored into Nevada Solar One's pricing. As the industry scales up over the next three to five years, she predicts, costs will fall by as much as 25 percent. Currently, the price of the electricity produced at Nevada Solar One is somewhere between 9 and 17 cents per kilowatt-hour. (Ausra, says Le Lièvre, is aiming for 10 cents.)

A second formidable competitor rose from the ashes of Luz. In 2004 Luz founder Arnold Goldman reassembled much of his old team as BrightSource Energy, based in Oakland, California. He abandoned troughs, convinced they would never compete on cost, and turned instead to another technology first explored in California in the 1980s: "distributed power towers." Instead of long, curved mirrors running horizontally across the landscape, Bright-Source will build its solar field using thousands of flat mirrors, seven meters square, made of standard glass. Each heliostat will be individually installed on an upright frame. Each will also have its own software controls, allowing it to track both the elevation and the azimuth of the sun; the aiming points will be calculated and updated every fifteen minutes. Sets of fifteen to twenty thousand mirrors will encircle and face seventy-meter concrete towers, each topped by a single receiver. Each tower, with its entourage of mirrors, will produce 20 to 40 megawatts.

By concentrating the solar radiation on a central receiver, BrightSource hopes to achieve higher efficiencies than Acciona's or Ausra's systems can attain: capturing more of the sun's energy as heat, turning more of that heat to electricity, and—by not pumping fluids through miles and miles of pipe—reducing the "parasitic load," the energy consumed in running the plant itself. Steady high-temperature, high-pressure steam will drive highly efficient supercritical turbines and convert up to 20 percent of the solar energy into electricity, double the efficiency of the original Luz troughs.

Goldman thinks utilities ultimately may get the most from their infrastructure investment by using the plants as "hybrids." Like a car that can switch between its battery and its combustion engine, the BrightSource plants could switch between gas and solar, allowing them to operate day and night and all year long. For now, though, utilities want "pure play" solar systems to meet mandates for renewable energy, so BrightSource is limiting gas use to "brief periods of passing clouds or cold mornings." In late August 2007, BrightSource filed an application with the California Energy Commission to build three plants with a collective capacity of 400 megawatts near the Ivanpah Dry Lake about sixty miles south of Las Vegas, a total of ten towers spread over eight square miles. With $50 million in venture funding, the company's Israeli subsidiary (Luz II) is developing its first pilot in the Negev, Israel's southern desert.

AN ANALYSIS BY SARGENT & LUNDY, an independent engineering firm, forecasts that by 2020 towers will produce the lowest-cost solar thermal electricity, down to 5.5 cents per kilowatt-hour, but as yet, no one has successfully pushed a tower receiver to the superhot temperatures BrightSource seeks. Abengoa, a $4 billion Spanish company, far outspends the U.S. Department of Energy on solar research, according to Fred Morse, a veteran solar industry consultant who is now helping Abengoa establish a foothold in the United States. In 2007 the company began operating the world's first commercial solar tower on a picturesque plain outside Seville in Spain. The graceless concrete monolith, 115 meters tall, is surrounded by 624 parabolic heliostats, each 120 meters square, whose curved mirrors will cost more than BrightSource's flat ones but will concentrate the sun more efficiently. Like the apes in the opening scene of the movie *2001: A Space Odyssey*, all the mirrors face the obelisk; shimmers of heat and light rising from the mirrors add to the otherworldly feel. The tower was built to work with molten salt at 600°C (1,100°F), but Abengoa has been operating it with steam at 250°C (480°F) to reduce fatigue on the components and ensure the system remains robust. The high-temperature tower-top receivers they had hoped to use remain unproved.

The Ausra team has tremendous respect for Goldman. "Bright-Source is a great competitor," says Le Lièvre. "They have a strong management team with huge experience. And their theoretical argument is spot-on—high temperatures are the gateway to higher efficiencies in power production and to better storage. If Goldman succeeds, there's no question he will have a low-cost technology." John O'Donnell, who is now Ausra's executive vice president, is most impressed by Goldman's next-generation technology, which BrightSource says will be ready to deploy in 2014. With new receivers (yet to be invented) and optics capable of multiplying the sun's power 1,500 times, BrightSource says it will heat compressed air to 1,200°C (2,190°F) and drive a "combined-cycle" plant, using exhaust heat to drive a second turbine. That would push efficiency to 24 percent. To get there, the company will need highly precise optics and positioning in its collector fields, exotic metals and ceramics tolerant of such high temperatures, and breakthroughs in fluid dynamics. "Some of the best work in the world on that topic has been done in Israel," says O'Donnell, "and I look forward to seeing how it comes out."

Ausra itself is building a research facility and experimental site in northern Germany where high-temperature technology will be explored; in 2003 the company built its first megawatt in "mini-towers," for which it has patents going back to 1995. "We know it well, and could change tack," concedes Le Lièvre. "But we still like low tech—low concentration ratios, low temperatures, and lower efficiencies—because we believe that in the end the cost per kilowatt-hour is lower." Le Lièvre points out that even though BrightSource has a strong track record, it is using technology that is new to the company. "Goldman is writing business at a low price, even though he's deployed just a fraction of a megawatt," he says.

"You can see the mountain he has to climb, and how many places they might stumble. But it's a real question mark. In ten years we'll see who was right."

Consultant Fred Morse is equally skeptical about Ausra. "[Vinod] Khosla is a marvelous spokesman," he says, "but they have never built a power plant, and until you build something, you can say anything you like." Based on his own experience, Morse, like Goldman, doubts that Ausra can fulfill its own promises—but says he wishes the company luck, noting that ultimately the market will sort out these competing claims.

THE THIRD MAIN STREAM of solar thermal technology development combines a mirrored dish with a Stirling engine. The most visible company is Stirling Energy Systems (SES), based in Phoenix, Arizona, and its core technology is the oldest of all. A Scottish minister invented the Stirling engine in 1816 (he called it "the Economiser"). The great nineteenth-century physicist and engineer William Thomson, Baron Kelvin of Largs, used one of the working models in his university classes, and Professor McQuorne Rankine (whose wonderful name now attaches to another kind of engine, which you will encounter in Chapter 7) fully explained the engine's elegant dynamics in 1850.

A Stirling engine, which can be powered by any external heat source, consists of a closed canister filled with a gas, which is transferred back and forth between the two ends, one hot and one cold. When the gas is in the hot end, it expands and drives a piston, crankshaft, and drive-shaft assembly much like those found in internal combustion engines, but without exploding the gas. The drive shaft turns a small electricity generator. The gas is then shifted to the cool end and recompressed. It has the highest efficiency of any heat engine—much higher, for instance, than an automobile's internal combustion engine.

In the 1980s, CEO Bruce Osborn began investigating the technology as part of a team at Ford Motor Company. In 1996 he cofounded SES with a holistic health entrepreneur named David Slawson, bought patents originally filed by McDonnell Douglas and six of McDonnell Douglas's 1980s test models, and signed a licensing agreement with Kockums, a Swedish defense company, to make and sell the engines.

By 2007, SES was testing six prototypes at Sandia National Laboratories near Albuquerque, New Mexico. Each unit has its own concentrator: a shiny dish, thirty-seven feet in diameter and made of seventy-eight mirrors, that tracks the sun. Each unit also has its own engine, held at the center of the dish by a fifteen-foot-long metal arm, bent at the elbow. At the hot end of the engine, facing the mirrors, the concentrated sunbeams are hot enough to melt lead. A time-lapse photo of the dish in action is oddly endearing, almost Chaplinesque. As the day dawns, the dish wakes to eagerly follow the sun across the sky; as darkness falls, it slumps in sleep.

In the summer of 2007, SES seemed to pull ahead in the solar thermal energy race when it signed power purchase agreements (PPAs) to supply 900 megawatts of electricity to San Diego Gas and Electric and 850 megawatts to Southern California Edison. Since each dish-engine combination generates just 25 kilowatts, those agreements will require SES to build and install seventy thousand dishes and seventy thousand engines. An artist's rendering of the vast fields SES plans to build looks like a futurist vision from the early twentieth century: an advancing army of parabolic bots, each with a big, solitary eye. Osborn believes he can realize that vision using mass-production techniques he learned at Ford. For instance, he says the support system for each unit is similar to a car chassis. He also sees the utility agreements as a useful lever for getting the things built: "We'll take those PPAs to suppliers and say, 'These are the cost targets we need to achieve.' " In late 2007, most analysts remained skeptical, noting that the company was way behind schedule, had yet to complete environmental reviews or file for permits from the California Energy Commission (CEC), was struggling to get investors, and continued to face questions about costs and reliability.

"Companies go public with these big utility deals to attract attention and enhance their value," says Le Lièvre. "The big numbers are useful to get people to focus on how significant solar thermal could be, and how quickly it can go on line if everything is favorable. But this is a wholesale business. I could ring all my customers in the space of one hour. If we make grand public promises and then fail, that affects those relationships. Utilities are all about 100 percent underwritten guarantees. For the industry as a whole, I don't think overpromising and underdelivering is useful." Not surprisingly, many in the industry complain that Ausra itself may turn out to be guilty of precisely that sin.

Negotiating a deal with a utility, it turns out, is just the first step, and lately a relatively easy one. Utilities like big, centralized plants, are at home with the steam cycle, and have become accustomed to large-scale renewable-energy contracts, thanks to the rapid development of wind farms. FPL Energy, one of the savvier utilities using solar thermal energy, has a pipeline of wind projects capable of generating 1.4 gigawatts and proposes to spend $650 million on high-voltage lines. Texas, another state rolling out the welcome mat for solar farms, has 3.3 gigawatts of wind power, and will have 4 gigawatts more if T. Boone Pickens goes through with plans to spend $6 billion on wind farms and transmission lines.

Utilities also see in solar thermal energy a hedge against an expanding array of financial uncertainties: the volatile price of natural gas, the rising cost of importing electricity from other states, and the expected cost of carbon. "Utilities are prepared to pay more to hedge their commercial risks," says Le Lièvre. "It's like a derivative, a financial instrument against the risks they face. They're buying this stuff because it adds real value to their portfolio mix. It's not about saving the planet but about doing what they're chartered to do—provide least-cost power."

Many utilities have a statutory obligation to use solar power. "Renewable portfolio standards" now in place in more than half the United States require utilities to buy a set percentage of their total power from clean sources; some specify that a fraction must come from solar. But this approach shares the same basic flaw as the European feed-in tariffs mentioned in the last chapter. It focuses on the wrong thing: on process (how utilities generate power) rather than on *performance* (how much pollution they emit), which is what ultimately counts. In essence, renewable standards, subsidies, and other mandates assume that the government has all the answers, rather than letting the market figure out the best way to produce clean energy at the lowest cost.

And if the government doesn't have the answers—if it picks the wrong approach—the results can be astronomically costly for consumers. Jay Apt of the Carnegie Mellon Electricity Industry Center estimates that at current economics, the cost of reducing emissions by generating electricity using photovoltaics rather than fossil fuels amounts to $400 to $650 per ton of carbon dioxide—ten times what the study predicts it would cost to capture and sequester carbon dioxide emitted by a coal-fired power plant, and nearly a hundred times the estimated cost of reducing carbon dioxide through energy conservation. Of course, entrepreneurs and venture capitalists like O'Donnell and Le Lièvre are betting that they can drive down the costs of solar-generated electricity (especially solar thermal electricity) far enough to be cost competitive. The power of a cap-and-trade system is that the government does not need to know any of this. Rather, it establishes the legal limits on pollution and lets the market do the rest. In John O'Donnell's words, "It creates specific must-do targets for emissions reductions, but leaves the choice of technology to the competitive dynamics of the marketplace."

In 2006 Governor Arnold Schwarzenegger signed into law AB32, setting the nation's first economy-wide cap on carbon emissions. It requires a 25 percent cut from 2006 levels by 2020, returning emissions to 1990 levels. "California offers outstanding opportunities to sell energy by virtue of its AB32 legislation," says O'Donnell. "Meeting the cap will require 17 gigawatts of new clean energy, about $50 billion of investment."* A second law adds the requirement that out-of-state power purchases must meet the same limits on global warming pollution as purchases in-state, which is having cascading effects throughout the American West.

The power of a cap on carbon in overcoming industry inertia became clear very soon after those laws passed. "Arizona Public Service wouldn't do a deal with us because our technology risks were still too high, which doesn't give us a chance to succeed," says Ausra's Le Lièvre. "But PG&E is writing business to almost everyone who steps up. They understand that you don't get to 10 cents a kilowatt-hour without reaching—we need partners willing to take the early risks. And once utilities move, they are very heavy, bring a lot of weight to the table and have a lot of money." As if to prove his point, in September, FPL Group (parent of FPL Energy) announced plans to build a $1.5 billion, 300-megawatt facility using Ausra's technology, "as long as Ausra agrees to FPL's cost and technical requirements and if the utility secures all necessary permits." A month later, PG&E confirmed a twenty-year agreement to purchase power from a 177-megawatt plant Ausra plans to build near San Luis Obispo, California.

---

* As of late 2007, California had just under 5 gigawatts of renewable generating capacity.

Almost all of these solar players now endorse a federal carbon cap-and-trade system. "A cap creates an immovable object, a problem that must be solved, and gets new technologies—first wind, now geothermal and solar—into volume production in the market where they then become 'ordinary' choices in resource plans," says O'Donnell. "A cap creates a market where companies can plan on sustained business and invest in aggressive cost reductions to be the folks who capture that market—exactly what the planet needs."

A cap-and-trade system will also provide the long-term predictability needed to fuel investment. "Regulations have to fit with the time frame of the industry," says Arnold Goldman of BrightSource. "Investors have to know that if it's going to take five years to site, get permits, and construct a plant, the economic environment will work five years hence. We're building something we hope to operate on a meaningful scale for thirty to forty years, so we need predictability over the long term." That predictability is exactly what declining legal limits on global warming pollution provide.

Venture capitalist Vinod Khosla explains the attraction of a cap on carbon emissions, how it would affect the "capitalist forces" that will decide the solar industry's fate, with his usual brazen flair: "A federal cap will increase the price of baseload power coal to about 8 cents per kilowatt-hour." At that point, he says, "solar kills coal." Perhaps a better way to think about the eventual mix of power sources is that solar energy is getting cheaper, while the price of using fossil fuels will inevitably rise as its true costs are built into the system through a carbon cap. A cap, in short, would help flatten the biggest obstacle for these start-ups: getting the hundreds of millions of dollars they need to build their power plants and fulfill their agreements.

"For companies that haven't yet deployed their technology anywhere, the question is how they'll attract construction finance," says Le Lièvre. "A big power deal with a utility is a great financial instrument, but only if you've priced it high enough to carry both the commercial costs and the risks of your project. Risk costs money. If you've got a high risk, you need to demonstrate good return on investment, to get people to park money in your projects. If you price your electricity too cheaply to make a good return and still have high technology risk, then it doesn't matter how many gigawatts you've promised to sell. It becomes worthless from a bank's point of view."

Ausra is working on several fronts to persuade banks to provide the $500 million it needs to build its first commercial-scale project. Its purchase agreements with PG&E, says Le Lièvre, provide "twenty years' guaranteed revenue, a good basis for debt financing. But how can a bank know we'll deliver on that agreement? We can get the turbine makers to put their balance sheets behind the supply and performance of the power block. But the concentrator piece, which is about half the cost, has only been operating since 2004. So we have to attract subcontractors and suppliers prepared to share this risk and put their balance sheet as backing. We use their company size to add credibility that we can deliver on time and budget.

"Still, there's an entire system risk that we as the integrator must carry. So we're also seeking more flexible bank financing—getting their engineers to understand how simple our technology is and how little can go wrong. We're finally starting to get strong competition to finance those plants out."

A number of U.S. banks have recently made big public commitments to renewable energy, inspired, in large part, by the growing expectation of a national carbon cap. Citigroup has pledged to invest $50 billion to address global climate change. (By 2007 it had committed $7.5 billion to renewable-energy projects, and over the next ten years it will invest more than $30 billion in clean energy and alternative technology. It will also spend $10 billion to reduce its own corporate environmental footprint, through its real estate portfolio, procurement, and energy use at its fifteen thousand global facilities.) Bank of America has committed $18 billion in lending, advice, and market creation to finance commercial production of new green products and technologies. At GE Energy Financial Services, global investing in wind, solar, biomass, and geothermal energy grew from $630 million in 2004 to $2 billion in 2007; by 2010 renewable-energy assets will make up 20 percent of the investment portfolio, $4 billion. JPMorgan has established an alternative-energy investment banking unit. Goldman Sachs, according to the *Financial Times*, is among the largest principal investors in clean technologies.*

To bolster the confidence of potential backers, Ausra will build a demonstration project in California. The company is negotiating a lease on a 5-megawatt power plant near Bakersfield that was originally built for experiments with zero-emissions natural gas combustion. It plans to convert the plant to solar technology, if necessary adding about 10 percent natural gas to superheat the steam to drive the existing turbine. It is also building a 6.5-megawatt plant in Portugal, planned to come on line in summer 2009 and run a small, custom-built version of the saturated steam turbines that Ausra will use on a larger scale (the smallest off-the-shelf saturated steam turbines generate 240 megawatts). By late 2008 Ausra hopes to have completed the eighteen-month permitting process with the California Energy Commission for its full-scale project, the 177-megawatt plant near San Luis Obispo. The company will likely stage construction of that plant, building just 30 percent of the solar field to start "so that the sharp edge of the technology bet will be $150 million," says Le Lièvre. Though they'd prefer debt

---

* *Financial Times*, May 24, 2007.

financing, "we could also argue in equity markets," he says, "and this time next year we will if the banks haven't come through. If we get both, we'll just grow more quickly."

In a June 2007 speech to the Western Governors' Association, John O'Donnell called financing "the last big obstacle to large-scale renewable energy deployment." While Abengoa built its solar tower in Seville with 90 percent debt finance at 3.5 percent interest, he told the governors, U.S. solar companies scramble to get 30 percent debt financing at 7.5 percent interest, and equity investors want 20 percent returns. "When Xcel financed the Comanche 3 coal-fired plant in Colorado," said O'Donnell, "they got 80 to 90 percent debt at 5.5 to 6 percent, and equity guys looking for 11 percent returns. If we could finance our plant with those terms, we could provide power at under 8 cents per kilowatt-hour. Then we'd be competitive not only with gas but also with pulverized coal. How fast renewable energy becomes an export crop for Western states depends on how fast we deal with risk protection and the risk-adjusted cost of capital."

THE OTHER MAJOR REMAINING OBSTACLE to large-scale centralized production of solar thermal electricity is effectively integrating it into the regional transmission grid. The first problem is access. The best sites for large-scale solar thermal plants are in the middle of nowhere, exactly where big transmission lines are not. Substantial new investment in the interstate transmission grid will be needed to bring the power generated by solar thermal plants from the desert to demand centers hundreds of miles away.

The second problem is cost. It is expensive to build major new transmission lines (more than $1 million per mile) and it is hard to earn a decent return on this kind of investment unless the line is used constantly. Because solar is intermittent and reaches maximum output only in peak sunlight conditions, a transmission line dedicated to serving solar thermal power plants will often sit idle or underused, like an empty toll road, failing to collect revenue for its owner. Which means either that the owner must charge a lot for the use that does occur or that he won't build the transmission lines at all. That is why new transmission lines for distantly located wind- and solar-generated electricity are so hard to get built, and why it is so expensive when they are.

The third issue is reliability. Managers of regional transmission grids worry that too much unpredictable supply connected to their system will make maintaining the crucial balance between supply and demand impossible, especially since demand is already hard to predict. There is much debate within the utility industry as to the magnitude of this challenge, but few deny it is there and needs to be addressed.

Viable storage systems would help overcome many of these challenges. David Mills of Ausra calculates that with the kind of hot water storage tanks he and his colleagues propose adding to Ausra's power plants, solar thermal energy could meet 96 percent of U.S. electricity needs. Such a high percentage will be possible, he argues, because of the close seasonal match between supply and demand (solar energy is most plentiful precisely when electricity is most needed for cooling; in winter, the main need is for heating, which is mostly provided not by electricity but by oil and natural gas).

An alternative strategy is being championed by David Olsen, the former CEO of the Patagonia clothing company, now a director at the Center for Energy Efficiency and Renewable Technologies in Ventura, California. In his "portfolio" model, utilities would fill transmission lines with power from a wide variety of renewable energy sources—wind, solar, biomass, and others—drawn from diverse geographic regions, thus greatly reducing intermittency and, in the process, reducing transmission costs. The renewable sources would provide the main power; existing fossil fuel plants would fill in when needed, turning today's model—in which renewables are the gap-fillers—on its head.

At the moment, however, says Le Lièvre, the California grid remains "a basket case" for renewable-energy power plants trying to get on. The U.S. Bureau of Land Management has received fifty applications for plants in the Mojave Desert, accounting for 50 gigawatts. "If you try to build out there alongside the Kramer Junction [formerly Luz] plants, you've got several gigawatts of potential projects queued up in front of you on the grid," says Le Lièvre. "You need to pay fees to the grid operator to upgrade to a level that will not only accommodate your load but all those that went before. There are dozens of gigawatts of plants that will never be built; they're basically squatting. Some of them are trying to get us to buy their place on line. You can imagine prices are stunning, but whether they're actually transferable is unclear."

California is trying to remedy matters. In April 2007 the Federal Energy Regulatory Commission (FERC) approved California's proposal to revise the rules for financing transmission lines for renewable-energy plants. Instead of requiring project developers to bear the full costs of interconnection, utilities can now build the lines and recover some of the costs from ratepayers.

Neighboring states are also overcoming their ambivalence about satisfying California's immense energy appetite. When in May 2007 the Arizona Corporation Commission turned down an expansion of a transmission line, one commissioner explained that "we don't want Arizona to become an energy farm for California." Until now, selling energy to California has meant mining coal, burning coal, and disposing of tailings, at great cost to western landscapes, air quality, and public health.

But farming solar energy is a different matter. The southwestern states have the best solar resources in the United States—among the best in the world. A March 2007 study for the Department of Energy found that Arizona, New Mexico, California, Nevada, Utah, Colorado, and Texas are the top states, in that order, and together have the resource and land to make more than 16,000 terawatt-hours per year from solar energy (total U.S. electricity generation in 2007 was about 4,000 terawatt-hours). In California, solar energy systems covering a 30-by-30-mile square could make 300 terawatt-hours, enough to supply the entire state's needs; 11 percent of New Mexico's land could be used to power the entire U.S. grid. The resource is particularly abundant close to big and fast-growing cities, including San Diego, Los Angeles, Phoenix, Las Vegas, and Denver. (Southern Europe and North Africa have equally valuable solar resources: 3 percent of Morocco could power the entire European grid. China, too, has rich solar resources, with some of the largest deserts in the world; in 2007 it committed to invest $200 billion in utility-scale solar power.)

In terms of the amount of land required, solar power plants are more efficient than wind farms, which need about six times the acreage per watt produced. Solar energy is also much more efficient than biofuels, which even with improved feedstocks and conversion processes will need more than thirty times the land per watt—although liquid fuels can do many things solar energy can not, an issue Chapter 4 explores.

The economic benefits of solar farming are also impressive. A 2006 study by engineering consultant firm Black & Veatch found that solar thermal plants create twice as many jobs as coal and gas plants and produce eight times the retained revenues in the states in which they are located. Each gigawatt of solar thermal–generated electricity, according to the National Renewable Energy Lab, will create 3,400 construction jobs, 250 permanent jobs, and $500 million in tax revenues.

America's sunniest states are now competing to create the most favorable policies. New Mexico wants to develop power plants that can produce 7 gigawatts of clean energy, half of it for export. To finance construction of the necessary transmission lines, the state has established a Renewable Energy Transmission Authority with the power to issue bonds. Colorado established a similar Clean Energy Development Authority, allocating $105 million to underwrite bonds for project development and construction of transmission lines. Texas is designating renewable-energy development zones and establishing public-private partnerships to finance grid upgrades to link these zones to population centers. "Our dream," says Le Lièvre, "is a solar park with grid access, permitting, and fabrication capability on a single site. Then we'd start construction and keep going for decades."

Not all grid upgrades will have positive impacts. In 2006 the California Independent System Operators, the group that runs the grid, proposed building a $1.3 billion, 150-mile transmission line, the "Sunrise Powerlink," to carry energy from the Imperial Valley deserts to San Diego. The line was intended in part to serve the first phase of Stirling Energy Systems' 900-megawatt project for San Diego Gas & Electric, but would pass through environmentally sensitive areas. The Center for Biological Diversity and other groups opposed its construction, calling the Stirling project "a technological Trojan horse," according to *Business 2.0*, "being used by the utility to justify an environmentally damaging big power grab that could ultimately be used to deliver fossil-fueled electricity."

On the federal level, Environmental Defense Fund is one of many organizations objecting to Department of Energy efforts to designate large swaths of the nation as "national interest electric transmission corridors" without conducting a programmatic environmental impact statement (PEIS). The corridor designation will enable FERC to use the power of eminent domain to site large interstate transmission projects, most of which will support new coal capacity. A PEIS would force the Department of Energy to consider better energy-supply alternatives and would lead to corridors more tightly tied to renewable-energy projects. Increasingly, advocates for wild land and wildlife are working collaboratively with clean-energy advocates, regulators, and grid developers to facilitate development of additional renewable transmission capacity while also protecting fragile ecosystems. In September 2007 the Western Governors' Association hosted a summit meeting on increasing the amount of electricity in the grid from renewable sources that drew representatives from two hundred organizations.

Ausra has joined forces with others in the solar thermal industry to press for a national network of high-voltage direct current (HVDC) lines, which have low energy losses over long distances (and are cheaper than alternating current lines for distances over five hundred miles) and could move power from solar farms in the Southwest deserts to the Midwest and Northeast for about 1.5 cents per kilowatt-hour, according to the Energy Department. Vinod Khosla estimates that to increase coal capacity enough to meet growing electricity demand would require a $20 billion investment in railroads. Investing that same money in a national grid, he argues, would provide greater stability. It would also enable renewable sources to satisfy a bigger proportion of the nation's total energy needs by taking advantage of regional differences in weather: excess supply in one sunny or windy state could meet excess demand in another. Europe is considering a continental grid to make use of its own diversity of resources. Solar energy produced in Spain during the summer could be shipped to Norway and used to pump water up into reservoirs at Norway's many hydroelectric plants; in the winter that water could be released through the turbines, powering all of Europe for several weeks.

In the meantime, Ausra is avoiding congested transmission lines by going for second-tier sites that have less intense solar radiation

and insufficient water for cooling. (The company will use less efficient dry cooling instead.) Ausra is also looking to build near existing wind facilities, to exploit underutilized portions of the grid. "In
California, it's rarely windy on stinking hot summer days, when
we'll operate at our peak," explains Le Lièvre. "Transmission lines
are rated for the peak load that a wind farm might produce, but
they almost never operate at that load. In Texas, wind farms use just
8 percent of their grid capacity, because it's often blowing at the
wrong time. So we can get a toehold in markets capturing capacity
in the grid that nobody thinks is there."

Ausra's 177-megawatt project will cover a square mile on a
ranch east of San Luis Obispo. The company leased the land from
Alberta Lewis, a fourth-generation Californian who was born in
1928 in Salinas. As a girl, Lewis helped her father drive cattle from
the train station in McKettrick over the Temblor Range to Painted
Rock, now part of the Carrizo Plain National Monument, which
is thirty miles from the ranch. In 2007 she was named San Luis
Obispo Cattlewoman of the Year; she is also an active member of
the Republican Women's Club and Farm Bureau. Three of Alberta
Lewis's four children now work the thousand-acre ranch. Daughter Susan Cochrane explains that the income from Ausra will help
them hang on to the property despite the estate taxes incurred by
the death of their father, Robert Lewis, in 2005.

As with all these technologies, the development of large-scale
solar power plants will require attention to other environmental values. The Carrizo Plain is one of the largest remnants of the
grassland habitat once abundant in the southern San Joaquin Valley. The Nature Conservancy calls it "California's Serengeti." In
1988 the Conservancy bought 82,000 acres and later handed them
over to the U.S. Bureau of Land Management and the California
Department of Fish and Game, which reintroduced native tule elk
and pronghorn antelope. By 2001 the protected area had grown to

250,000 acres and President Clinton had designated it a national monument. More endangered species, including the San Joaquin kit fox, blunt-nosed leopard lizard, giant kangaroo rat, California jewelflower, Lost Hills saltbrush, Kern mallow, and San Joaquin wooly threads, live here than anywhere else in California. The San Andreas Fault runs like a visible scar across the edge of the plain. In winter, lesser and greater sandhill cranes take shelter in Soda Lake, one of the largest undisturbed alkali wetlands in the state. At Painted Rock, pictographs created around 2000 BC by the Chumash people are preserved in an alcove of sandstone. The Sierra Club finds on the Carrizo "a wildness on a scale that allows us to imagine what California was like 300 years ago."

Ausra's chief development officer, Rob Morgan, who spent fourteen years at AES, one of the world's biggest power companies, is in charge of managing environmental impacts. "We have no emissions and no wet cooling," he says, "which eliminates right off the two biggest impacts of power production. But we are constructing a major piece of equipment in untouched desert areas. The core concern is species: making sure that the habitat we pick isn't supersensitive. We've spent $10 million on flora and fauna studies and developed plans to mitigate problems, for instance, by relocating fauna. We're working with U.S. Fish and Wildlife, state fish and game and local environmental groups. And we'll have to pass the same stringent tests as a coal plant to get permits from the California Energy Commission." As with all power projects, however, matters of scale will be highly important. Tom Malone of the Nature Conservancy in San Luis Obispo says that while a single square-mile solar installation will have negligible impact on the surrounding ecosystem, "we wouldn't want to see the whole valley turned into a solar farm." In the Carrizo, a proliferation of large solar facilities might close off corridors used by the kit fox.

Susan Cochrane, who in five decades of herding cattle has

learned the names of most of the grasses and trees that survive the 100°F summertime heat and persistent drought, and has watched rabbits and burrowing owls take shelter in the shade of her watering troughs, is not worried. In the 1980s, she says, there was another "mirror farm" in just about the same place where Ausra's project will be, although no one now remembers who set it up. "It never bothered me," she says. "It looked just like a mirage, just like you always see out there." Although one new neighbor, who recently bought forty acres, worries that the new installation will affect the resale value of his land, Susan thinks she will prefer it to the abandoned campers and broken-down mobile homes that now mar her view. She also thinks Ausra's plant will create some good jobs.

And like many, she simply wants to believe that solar energy's time has come—or that it will come sooner rather than later. "It's natural and totally renewable," she says. "We have three daughters to think about, and their future kids."

CHAPTER 4

Fuels from Living Creatures

I n the spring of 2007 Amyris Biotechnologies in Emeryville, on the eastern edge of San Francisco Bay, still bore the marks of a hot little start-up. There was the foosball table, and on the floor beneath it the masking tape marking the spot where new lab benches would soon go in. There was the senior vice president of research, Jack Newman, looking like he was just out of college in worn jeans with hair to his shoulders. There was the antic spirit: asked if he was ever tempted to drink the deliciously yeasty-smelling potion bubbling away in nearby fermenters, Newman admitted that he and his colleagues tested each new instrument by brewing beer in it first. Most of all, there was the buzz, the crackling energy that comes from the mix of youth, world-changing ideas, and lots and lots of money. The most famous venture capitalists in Silicon Valley had recently put $20 million into the firm, and its new CEO promised to grow it into a $10 billion company in five years. It's what you might imagine it felt like when Apple was just a seedling, or when Andrews, Clark & Co. was born in 1863—soon to become Standard Oil.

Here's why Amyris has had to turn away investors: Using a platform they developed when they were postdoctoral students at University of California, Berkeley, the founding scientists have reengineered the metabolism of yeast to ferment sugar into a pure hydrocarbon fuel. Unlike the ethanol usually made from sugar, this fuel is as energy dense as gasoline, and can be shipped through existing pipelines and pumped into any car now on the road. The young scientists also have engineered "plug-and-play" genetic modifications they can pop into their yeasts to make them produce, from sugar, everything we now get from a barrel of oil, ranging from industrial chemicals and plastics to diesel and jet fuel.

BIOFUELS MAY PLAY A SMALLER ROLE than photovoltaics or solar thermal electricity in reducing global carbon emissions, for one simple reason: plants are far less efficient at converting an acre of solar radiation to usable energy. They did not evolve, after all, to optimize energy production, but to adapt, survive, and reproduce. Even switchgrass, a cutting-edge energy crop, is less than one-hundredth as efficient as the best solar cell. It converts just 0.3 percent of incoming solar energy into chemical energy; Spectrolab's solar cells, by contrast, convert 42 percent. It also has ongoing needs for nutrients and requires all the work of growing, harvesting, and processing. The amount of water demanded by biofuels production is immense, with most crops requiring about a thousand tons of water for each ton of biomass.

There is also a notable irony associated with biofuels production. In pursuit of climate-friendly transport fuel, the industry has generated increased demand for coal—the most polluting of the fossil fuels. Throughout the Midwest and Plains states, ethanol developers either are building their own coal-fired boilers to generate the heat and pressure they need, or are buying electricity from local utilities or co-ops that, in turn, are planning new coal-fired power plants to meet the new demand.

In a civilization as centered as ours on the automobile, however, and a global economy so dependent on transport, fluid fuels will necessarily play a crucial part. While there are many clean ways to make electricity, and while electricity may in the future become the prime power source for transport (Chapter 9 explores this idea), there is simply no immediate substitute for liquid fuels, which as currently produced inflict a heavy burden: the U.S. vehicle fleet pumps 1.3 billion tons of carbon dioxide into the atmosphere every year, and $820 million in capital is exported every day for the oil needed to do so. The concentration of chemical energy in a gallon of diesel or gasoline and the ease of storing and delivering that energy are unmatched. As Caltech professor Nate Lewis says, "You can take a $5 piece of hose and pump at a rate of 10 megawatts into your car. To move 10 megawatts of electricity, you need high-voltage transmission lines. And the electricity won't just sit there, like the gas, waiting for you to be ready to use it." Nor is it just cars and planes that rely on petroleum: the $1.5 trillion chemical and plastics industries also depend on the stuff.

Brazil has already demonstrated that "you can change the paradigm," as Jack Newman says, "that petroleum is what you burn in your car." Using its abundant sugar crop, that tropical nation has replaced 40 percent of its gas with biofuels. (Fully 85 percent of the cars sold in Brazil are "flex-fuel," meaning that they can burn either ethanol or gasoline; most U.S. cars can burn only a small amount of ethanol mixed in with the gas.) "That's $59 billion since 1975 *not* shipped to the Middle East," says John Doerr, who is backing Amyris and sits on its board. "And a million new jobs." In 2006 Brazil became officially energy independent, its ethanol exports equal to its petroleum imports. (Changing the paradigm in the United States, of course, would demand a much more dramatic commitment; we produce about the same amount of ethanol as Brazil, but that ethanol displaces less than 1 percent of the gasoline we consume.)

Although they are challenging the most consolidated segment

of the energy business, ruled by famously recalcitrant titans like
ExxonMobil, the biofuels start-ups have much in their favor. The
global oil supply is increasingly problematic. As easily accessible
reserves mature, oil companies are forced to seek new supplies that
are more difficult and expensive to develop and often contain more
impurities, particularly sulfur; finding and developing these new
oil reservoirs, then refining the "cruder" crude they yield, means
rising prices at the gas pump. With fully 50 percent of the world's
oil reserves now owned or managed by state-controlled oil companies, it becomes even more difficult for the oil giants to replace
their reserves. Meanwhile, it becomes clearer every day that the
United States' overwhelming dependence on foreign oil is a direct
risk to national security, and that we must expand and diversify our
sources of energy.

Because they are so new, and still evolving, biofuels start-ups have
a further advantage in this era of rapidly changing technologies:
they can fully exploit ongoing advances in computing, biotechnology, genomics, and other sciences. Surveying his lab, Newman
gives up counting: "There are a hundred enabling technologies in
this room."

And by blurring the boundaries between energy, agriculture,
and biotechnology, these companies are generating all kinds of
unimagined inventions. Carlos Riva, who recently became CEO
of a company called Verenium, which is developing technology to
turn nonfood materials like grass and woodchips into "cellulosic
ethanol" (Chapter 5 takes a closer look at cellulosic feedstocks),
comes from a traditional energy background. From 1995 to 2003,
he ran InterGen, an independent power producer jointly created
by Shell and Bechtel which developed more than 17,000 megawatts of electric-generating capacity, along with gas storage facilities and pipelines, on six continents. His chief financial officer, John
McCarthy Jr., spent the last fifteen years in the healthcare/life sci-

ences industry. "We speak entirely different vocabularies and talk past each other sometimes," says Riva. "But there's such a richness of diverse business practice and experience, our perspectives are so different, it makes you jettison all your old thinking and invent something brand new." Companies like BP and Chevron, staffed for decades almost exclusively by geologists and engineers, are now scrambling to catch up, trolling graduate schools to hire bioengineers. Some of their top executives are abandoning ship entirely and crossing over to the biofuels side.

A FEW BIOFUELS START-UPS employ thermochemical refining techniques like their petroleum competitors. But most convert biomass into fuels using microbiological processes from beginning to end. This is not just a technology choice but also a kind of philosophy: most believe that it makes sense to tap the millions of years of research and development already completed by nature, building on biological processes that are typically less dependent than industrial processes on expensive energy inputs and harsh chemicals. Janine Benyus, author of the influential 1997 book *Biomimicry: Innovation Inspired by Nature*, describes what she calls the "master chemistry" of nature. To make Kevlar fibers, for instance, DuPont must heat petroleum to 760°C (1,400°F) and boil it in sulfuric acid. "Living organisms can't afford to heat things to high temperatures or do pressures or chemical baths," Benyus says. "The orb-weaver spider turns flies into a material five times stronger per weight than steel.* Silently, in water, at room temperature."

Relying on biology has its costs, however, including impacts on food supplies, water resources, biodiversity, and local economies as land is cleared to farm fuel. The United Nations predicts that global demand for both food and fuel will double by mid-century, and

---

* The spider's genetic makeup allows it to produce special enzymes that catalyze the process.

warns that a boom in biofuels could reduce food production and drive up prices, particularly in famine-ridden regions like Africa. In Brazil, where millions of citizens are poor and hungry, half the sugar crop already goes to make fuel.

The unintended consequences of the biofuels boom have been even more severe in the case of biodiesel made from palm oil. Several European countries were heavily subsidizing palm oil diesel to meet their greenhouse gas targets until they discovered that rainforests were being felled across southern Asia to meet that demand, producing a net *increase* in atmospheric carbon. Friends of the Earth estimates that 87 percent of the deforestation in Malaysia from 1985 to 2000 was caused by new palm oil plantations. In Indonesia, those plantations expanded tenfold over the past twenty years, in large measure through the draining and burning of peatlands. Peatlands, when intact, are among the world's greatest carbon "sinks"—naturally soaking carbon dioxide out of the atmosphere and sequestering it in the ground. So the expansion of Indonesia's plantations has had a decidedly adverse effect, releasing billions of tons of carbon stored over time back into the atmosphere each year.

The amount of land required to grow enough biomass to displace significant amounts of petroleum with biofuels is daunting. The United States uses 140 billion gallons of gasoline and 40 billion gallons of diesel fuel each year. Converting the nation's entire soy crop to biodiesel would meet just 6 percent of diesel demand— and would increase pressure on producers like Brazil, where rainforests continue to be cleared at a furious pace to plant soy to meet global demand. At current average yields, according to biomolecular engineering professor Kyriacos Zygourakis of Rice University, replacing just 30 percent of the gasoline consumed in the United States with ethanol made from switchgrass would require 200 million acres, equivalent to about half the total cropland in the United States.*

Land use issues have led some to argue that the conservation and restoration of forests and savannahs should precede any move toward biofuels. Like the peatlands of Indonesia, soils and vegetation around the planet store an enormous amount of carbon; when cleared for agriculture, that carbon is released into the atmosphere. In an August 2007 article in the journal *Science*, Renton Righelato of the World Land Trust and Dominick Spracklen of the University of Leeds calculated the amount of carbon released to the atmosphere when forest is cleared to plant crops for biofuels, then compared that to the amount of carbon saved by substituting those biofuels for gas and diesel. Even after thirty years, they concluded, the "up-front emissions cost" would exceed the emissions avoided by switching to alternative fuels. In fact, preserving and restoring forests and grasslands would sequester up to nine times more carbon, while also preserving biodiversity and reducing nutrient runoff and soil erosion. If landowners could then sell those carbon offsets in a global market, many—especially in rainforest nations—would come out well ahead, an idea discussed in Chapter 9.

Perversely, the current state of affairs not only fails to reward landowners for sequestering carbon but also subsidizes "solutions" that create more problems than they solve. Corn-based ethanol has proved particularly problematic on this count. For political reasons, it has captured the lion's share of federal subsidies in the United States and comprises more than 90 percent of the nation's biofuels production. But it has distinct drawbacks. It puts two vital human needs—energy and food—in competition with each other. And it is a relatively poor energy producer. To fill one 25-gallon tank

---

* From a speech at the October 2007 World Oil Conference in Houston.

with corn ethanol requires enough grain to feed one person for an entire year.

In 2006, 20 percent of the corn produced in the United States was converted to ethanol but it displaced just 3.5 percent of gasoline demand. The carbon reductions of corn ethanol are also negligible, given all the energy needed to plow the fields; grow, harvest, and transport the corn; and then turn it into fuel. According to a report for the National Academy of Sciences, ethanol producers get just 25 percent more energy out than the fossil energy they put in.[*] As a result, measured over its whole life-cycle, corn ethanol is not very good at reducing greenhouse gas emissions. Alex Farrell and Daniel Sperling of the University of California, who helped develop that state's low-carbon fuel standard, assign a number called "global warming impact" (GWI) to each fuel.[†] The GWI for gasoline is 92; corn ethanol, 76; Brazilian sugarcane-based ethanol, 36; and cellulosic ethanol, just 4. The nitrogen fertilizers used in industrial agriculture also boost emissions of nitrous oxide, a persistent and potent greenhouse gas. Switching to "no-till" farming, which keeps carbon stored in the soil and reduces the need for added fertilizer, helps the energy and carbon balance, but not enough to warrant a long-term commitment to grain ethanol.

The Organisation for Economic Cooperation and Development (OECD) has called for the end of all biofuels subsidies, arguing that a switch to these fuels would cut energy-related emissions by just 3 percent while exacting huge costs.[‡] With the United States alone spending $7 billion a year on ethanol, the OECD calculated, each ton of avoided carbon dioxide costs more than $500. "As long as environmental values are not adequately priced in the market,

---

[*] From *Proceedings of the National Academy of Sciences of the United States of America*, July 2006.

[†] GWI is measured as grams of carbon dioxide per megajoule of fuel burned.

[‡] From the OECD report, "Biofuels: Is the Cure Worse Than the Disease?" September 2007.

there will be powerful incentives to replace natural ecosystems such as forests, wetlands, and pasture with dedicated bio-energy crops," the report concluded. The OECD recommended that rather than rigging the market in favor of a particular technology, a "technology-neutral" price be put on carbon to allow the market to find the most efficient ways to reduce greenhouse gases. A technology-neutral price on carbon, of course, is exactly what a cap-and-trade program would create—while at the same time guaranteeing long-term cuts in emissions.

Biofuels innovators will have to navigate all of these tensions, and the start-ups are well aware of the challenge. Amyris fuels promise to relieve some of the ecosystem pressures: because sugarcane is much better than corn at "fixing" carbon (that is, using photosynthesis to turn atmospheric carbon dioxide into carbohydrates), and growing and harvesting it requires less input of fossil fuel (for fueling tractors and fertilizing), the net reductions in carbon emissions from Amyris fuels exceed 85 percent, around seven times those of corn ethanol.

GROWING UP IN NORTHERN CALIFORNIA in the 1970s, Jack Newman knew from an early age that he wanted to study "the chemistry of life, the most powerful of all sciences," and use that knowledge to change the world. "I was young enough not to be embarrassed by that idealistic vocabulary," he says now. He took a somewhat long way around to that destiny. First he dropped out of high school because he was bored. Then he enrolled in junior college, but dropped out again, this time to follow his first love, an aspiring chef, to San Francisco and Paris. In both cities he worked as a waiter ("I wouldn't mind doing it again," he says. "I'm a hedonist at heart"). He felt free to "do the vagabond thing," he says, "because I knew exactly what my future path would be." The couple eventually returned to Berkeley, where Newman completed his degree in biochemistry at the University of California. (They later parted ways, though remained friends: when Newman married in 2001, she baked his wedding cake.) He had one last footloose moment—working as cook and crew member on a sailboat in the Caribbean—before beginning graduate school at the University of Wisconsin in Madison. In 2000, Newman returned to Berkeley to do postdoctoral work in Professor Jay Keasling's laboratory; it was there that he met protein biologist Kinkead Reilling, then twenty-eight, and chemical engineer Neil Renninger, twenty-five, who would become his partners.

While earlier generations of genetic engineers had focused on single-gene transfers to produce single proteins (using recombinant *Escherichia coli*, for instance, to make human insulin), Keasling's team of researchers was working to modify multiple genes to work together to make a variety of complex molecules. By "hacking" a microbe's entire metabolic system, they could turn a bacterium or yeast into a living factory, able to do in one step what might take seven or eight steps involving intense heat and chemicals in a standard plastics plant or refinery. These microbial factories would be more efficient, more flexible, and far cleaner than conventional chemical plants. And they wouldn't need to be fed fossil fuels: they could start with simple, cheap sugar and end up with the stuff that powers the world.

As postdocs, the three young scientists built the necessary "platform" and wrote the new enzymatic software for the yeast. Newman and other key members of the team published their findings in the July 2003 issue of the journal *Nature Biotechnology*. Then—"over lots of good wine and bad Chinese food"—they set about figuring out what to do with their new living machinery.

The first answer came from the World Health Organization. In 2002 the WHO had made an urgent call to the medical community to cease treating malaria with quinine, to which the parasite was becoming resistant, and instead switch to an ancient Chinese remedy called artemisinin. Derived from sweet wormwood, artemisinin, in combination with other readily available drugs, is nearly 100 percent effective in curing malaria. The problem is that it is expensive and takes lots of land to produce: to grow the crop and extract the drug costs about $2.50 per cure, an order of magnitude too expensive for the developing world. As the young scientists watched the price spike, Newman recalls, "We thought, 'Maybe there's a company here.' It was incredibly high risk. We were going to rebuild the entire metabolism of a plant inside yeast, turning something that makes beer into something that makes a drug on an industrial scale—and we were going to do it much more cheaply than planting plants and extracting the drug. To a little postdoc, it was like, 'Are you kidding me?' But we knew if we could do this it would be proof that we could do anything."

They launched their company in 2003 and—because its mission was to replace one scarce, expensive source of a valuable commodity with another—named it after amyris, a plant that had replaced sandalwood when the supply of that fragrant wood shrank and the price soared. Their initial funding came from an unusual source. The Institute for OneWorld Health, a nonprofit drug development company, had just received $43 million from the Bill and Melinda Gates Foundation, and passed $12 million on to the Berkeley researchers. It was a scary experiment, OneWorld Health's founder, Victoria Hale, told a reporter. "We were the first venture capitalists for Amyris—the only ones."

That high-risk investment paid off. The Amyris technology has cut the cost of making 100 million doses of artemisinin from $250 million to less than $100 million, in effect earning the Gates Foundation a $150 million return on its philanthropic investment. In John Doerr's words, "Amyris leverages millions of years of evolution and will save a million lives a year."

It was time to start thinking about what to do next. "Believe me, if the only thing we ever did was cure malaria in the developing world, I'd die a happy man," says Newman. "But all along we kept trying to find fuel applications because we're all enviros." Their initial work in Keasling's lab had focused on bioremediation: creating microbes to eat toxic waste. "But we came to understand that the way to deal with pollution is not post facto, but 'pre facto'—by coming up with a process that moves you away from deep-sea drilling and pumping of oil, and toward fewer emissions and greater biodegradability."

The Amyris researchers dismissed corn-based ethanol as too flawed: its energy density is just 70 percent that of gasoline; it requires huge amounts of energy to separate the 10 percent of ethanol (the grain alcohol) produced from the 90 percent fermentation steep (the beer); and because it is miscible with water, and there is always some water in pipelines, it must be transported in fuel-burning trucks to avoid dilution. "If you put it in a pipe, by the time it gets from the Midwest where it's made to the coasts where it's used, it isn't even a martini," says Newman. "It's a gin and tonic."

The question was, Could they modify their microbial drug factories to make a new fuel, an ideal fuel, from scratch? "We looked at the *Merck Index* [the professional encyclopedia of chemicals, drugs, and biologicals] and said, 'If you could pick any molecule to use as fuel, what would you pick?'" says Newman. They selected several candidates, then set about designing the right metabolic pathways to make them.

Again, they began with sugar, either from cane (selling in 2007 at 7 cents a pound) or—better yet—from cellulosic materials such as straw and wood chips. (Other innovators are working to "crack" cellulosic materials with enzymes or thermochemical processes to obtain sugar at one-tenth the price.) Again, microbes (yeast or bacteria—Amyris researchers are working with both) would provide the conversion machinery. "Yeast catabolyzes sugar," explains Newman. "It extracts energy for its own use and excretes ethanol. What we've done is reroute the enzymatic steps to get it to excrete more of the energy—tricked the yeast so that it's in its own best interest to do so. It's what the ancient Greeks did when they sampled two different bottles of wine and chose to make more of the wine that got them drunker—they selectively bred yeast that produced more kick per unit of sugar. We're just synthesizing that process, and radically accelerating it.

"Essentially, we reprogram the yeast the way you'd reprogram a computer, and with the same increase in speed. Just a few years ago, to get an organism you wanted, you had to make millions of variations and try each. And it took six or seven years to do a single iteration. Now, I can do a targeted design of five promising genetic modifications, write code in the form of DNA, order it online, and have it by FedEx within a week. We make a strain, test it in a shake flask, then move it to a larger vessel and simulate the pressure of scaling up. Along the way we debug it, like an operating system. It runs and hangs up and we use one of our plug-and-play fixes to get it going again."

With their drug research and development nearly finished, in 2006 Amyris went looking for $7 million in venture capital to fund their new project in biofuels. In two months they had $20 million and were turning investors away. That first round was led by Khosla Ventures, Kleiner Perkins Caufield & Byers (KPCB), and Texas Pacific Group Ventures, all of which placed representatives on the Amyris board. Amyris hired a new CEO, John Melo, who had been president of BP's U.S. fuels operation.

By 2007, Amyris researchers had produced gasoline and diesel from sugar, powered a diesel truck engine, and begun emissions testing. Because they control all the inputs, says Neil Renninger, they can also control what is emitted when Amyris's crystal clear, mild-smelling fuels are burned. They emit none of the smog-forming combustion products that come from burning aldehydes like ethanol and butanol. And if any troubling emissions do emerge, Renninger claims, they can tweak the organism to take the bad stuff out. By 2010, says Melo, Amyris will have a market value of $10 billion.

They will—*if* they can cut costs enough to make a product cheap enough to burn. Though they will not say where they are in their walk down the cost curve, they claim to be ahead of schedule to get to market by 2010 at under $2 a gallon. Because their fuels have much lower net carbon emissions than petroleum-based fuels, a cap-and-trade system will significantly improve their competitive position. According to board member John Doerr, a cap on carbon will increase Amyris's profit potential by at least 25 percent—and thus also increase the incentive to get the new fuels to market.

Frances Arnold, a chemical engineering and biochemistry professor at Caltech and a member of the Amyris scientific advisory board, sees a challenging path ahead, for Amyris and for her own biofuels company, Gevo, which makes biobutanol, another energy-dense fuel. "None of us really knows yet how to lower costs. We engineer organisms to do our bidding, but they don't have high enough productivity. And switching out genes is like writing *Moby-Dick* by pulling discrete paragraphs off the Web; you end up with a kludgy design." Arnold's researchers have developed strategies for using evolution as an "editor" to smooth out any inelegant passages in the genomes they write. They design a genome for their desired organism, but then introduce thousands of random changes in the DNA. From those mutations, they select the best, then do it again. The approach is called "directed evolution" or "semi-rational design," because it mixes the high-speed, directionless shuffling of mutations with the conscious writing of genetic code.

As important as working out the technology, say the Amyris principals and their backers, will be getting the right state and federal policies in place. Newman was part of the group of entrepreneurs and venture capitalists, led by Doerr, who went to Sacramento to push for California's cap on carbon emissions. (It was the first time, says Newman—who worked for the League of Conservation Voters at age seventeen and later for Earth First!—that he had gone to the state capital without getting arrested.) The group told lawmakers that the best way, the only way, to find carbon reductions that are cheap, quick, and sufficiently far-reaching was to mobilize the market. That meant setting an upper limit on total carbon emissions and allowing emitters to trade their allotments. Once there were profits to be made from reducing pollution, capital would flow in and innovation would flourish.

After the bill passed, the group shifted its focus to the nation's capital, arguing for a federal cap on carbon emissions. To make the case in Washington, Amyris hired a former BP executive, Mike McAdams of Hart Downstream Energy Services, and sent its CEO to testify before the U.S. Senate on behalf of renewable fuel standards based, like California's, on the fuels' global warming impact.

BY LATE 2007, Amyris had grown to 120 people and moved to bigger quarters. The firm hired three more executives with long histories at BP, including senior vice president Paul Adams, who had overseen the oil company's supply and trading operations throughout North and South America. They began discussions with Costco about substituting Amryis biofuels for ethanol in Costco's blended fuels. And soon after the European Union announced that it would set carbon caps for commercial airlines by 2011, Virgin Fuels, an investment firm owned by Sir Richard Branson of Virgin Atlantic, approached Amyris about developing a low-carbon jet fuel. From the company's standpoint, the idea was very intriguing. Concentrating on jet and diesel fuel first, says Doerr, means Amyris can sell "to a smaller number of large commercial buyers to get to scale quickly." As with all of his firm's clean-tech investments, he says, "though these are tiny start-ups, we do have an agenda of speed and scale. We believe they will have measurable impacts within five years."

Jet fuel is particularly hard to make from biomass: it must have a freezing point low enough to withstand high-altitude temperatures and an energy density high enough to allow planes to fly long routes without added weight. "The fact that no one else was addressing the problem in a sustainable way focused us on the problem," says Amyris CEO Melo. "We realized we could make a big impact if we developed a fuel with zero-sum carbon emissions."

As they had with gasoline, the Amyris researchers began with the current standard fuel, called jet-A, and "asked ourselves if we could generate a fuel with more energy and a colder freezing point, which would enable flight over the poles," Melo explained in the June 11, 2007, issue of MIT's *Technology Review.* "We identified a molecule that we believed our core technology could make and then set out to design microbes to make that product. Now we've been able to make it efficiently enough that we believe it would allow us to make a jet-A equivalent with better properties on energy and freezing point with a $40 barrel [of oil] cost equivalent by 2010 or 2011."

Their long-shot success with drug synthesis continues to energize the players. "When we started the artemisinin project, our odds were a million to one. No one thought we could do it," says Newman. "Now five of the world's biggest pharmaceutical companies have voiced interest in producing the drug, knowing we want a commitment from them to distribute it at cost in Africa. [Amyris has agreed to make no profits on those sales.] To see that become a concrete reality has given everyone a sense that we can do the next hard thing, the fuels, in three years' time. The intensity of having a nearly done project is driving all our work forward."

For a company that displays images from Comedy Central's cartoon show *South Park* on its reception desk, identifying the alter egos of each of the three young founders, the influx of corporate heavyweights has required some cultural adjustment. "The lab rats in black that are all pierced and tattooed have had a harder time getting used to the guys in jackets than the other way around," Newman says. For his own part, he misses the "intimate" feel. "I used to make my mom's noodle salad for our company picnic, but this year we had to hire a caterer." He still manages to get in some surfing, sometimes even with a competitor: Steve del Cardayre, vice president for research at LS9, another synthetic biology company working on similar fuels.

As Amyris faces the possibility of becoming an energy behemoth in its own right, Newman wrestles with the implications of success. His favorite headline in the satirical *Onion* captures both the promise and the peril of their work: "Scientific Breakthrough Fixes Problems Caused by Last Scientific Breakthrough." He sees a potential for social benefits in the rebalancing of world demand for sugar and corn. A fan of Jared Diamond's *Guns, Germs, and Steel*, on how natural resources and climate shape national destinies, he believes that a shift from a global agriculture overwhelmingly dominated by corn for food to one that also values the fuel potential of sugarcane might affect the world's balance of power in favor of poorer, equatorial nations. (Newman's optimism contrasts sharply with the darker view of some groups, which warn that small farmers in poor countries could be turned into wage laborers on agrofuel farms owned by multinationals, and also fear environmental damage from monocropping and the use of pesticides and genetically modified organisms.)

Newman also sees a possibility for new kinds of leverage in the world. "If we're successful, we'll be wealthy, which means we'll be in a position to impact politics. That may be a massive rationalization. But part of my agenda has been to create a power base to change the world. The skill set I have to do that is technological, but I know in the end it's all about the money."

Amyris alone, Newman says, cannot possibly solve all the biofuels supply-chain problems. He and his colleagues are hoping for the success of other start-ups that are working to crack open cellulosic material, the world's cheapest and most abundant biomass, which is neither edible nor nearly so demanding of land, water, fertilizer, and energy. "Fortunately, people are investing tons of money right now in figuring out cellulosic technologies," says Newman. "When they get it done, we'll definitely plug it into our organisms."

CHAPTER 5

New Sources of Biofuels

Sugar is the very simplest biomass to turn into fuel: it can be fermented directly into the high-octane moonshine that is ethanol, or into the new, improved liquid fuels Amyris is designing. In Brazil sugar is converted to ethanol at a cost of 60 cents a gallon, getting 8 BTUs out for every 1 BTU put in. The second easiest plant material to convert is starch. A few cents' worth of enzymes turns the stuff inside a kernel of corn to sugar, and from there to ethanol, but with a far worse energy balance: just 1.3 BTUs for every BTU put in. The hard stuff is nearly all the rest: the tough, fibrous, "cellulosic" material that makes up grasses and stalks and husks and cobs and tree trunks and branches and leaves.

Plants have good reason to put most of their sugars into cellulose and hemicellulose. They need structure to grow toward the sun. And to deter potential predators, it is useful being difficult to digest. Cellulose, in fact, is the single most prevalent form of carbon in nature. It is well suited for biofuels precisely because it is not food—and because when it is turned into fuel, its energy balance is excellent: up to 36 BTUs for each BTU put in. But it is also the most difficult to use. The crystalline structure of cellulose— very long chains of six-carbon glucose molecules—makes it hard to dismantle. Breaking those strands requires special enzymes: proteins that catalyze a chemical reaction without being used up. In 2007 the enzymes required to turn cellulosic material into fuel cost more than 50 cents a gallon, about twenty times the price of the enzymes needed to convert corn. Hemicellulose is a more random and amorphous structure of connected sugars and is therefore easier to break down, but it contains some five-carbon sugars (primarily xylose) that no one, until recently, knew how to ferment. Plants also contain lignin, which provides further structural integrity and, with the energy density of coal, can be burned to power biofuels refineries.

No one has yet managed to build a commercial-scale plant in the United States capable of converting cellulose to ethanol, though companies are developing plants in Georgia, Florida, California, Iowa, and Idaho. In February 2007 a company called Verenium, based in Cambridge, Massachusetts, began construction of a 1.4-million-gallon demonstration plant in Jennings, Louisiana, that will make fuel from bagasse—the fiber left over after the juice has been squeezed from sugarcane. Their primary feedstock will be an "energy cane" related to an aboriginal species that grew in the southeastern United States two hundred years ago; those native cultivars grow like weeds on land unsuitable for other crops, and stay standing even when killed by frost, making harvest possible. The demonstration plant will also convert wood and perennial grasses to biofuels, producing up to 2,000 gallons from each acre harvested, five times the yield of grain-based ethanol. Achieving those yields with a nonfood crop that requires few inputs and is grown on marginal lands could completely change the energy and carbon economics of biofuels, particularly if combined with Amyris-style fermenting technologies.

At the groundbreaking ceremony for the Jennings plant, just eighteen months after Hurricane Katrina, Louisiana Governor

Kathleen Babineaux Blanco thanked Verenium for mapping a new direction for her battered state. "We will no longer be simply an oil and gas state, but the state that started the cellulosic ethanol revolution." With its subtropical climate and abundant rainfall, America's southeastern agricultural region—much of which is now in pasture or rice and making low margins—could be put to far more profitable use. Because feedstocks are bulky and transporting them is costly, biofuels producers try to build refineries adjacent to the fields. Verenium is locating plants so that sufficient biomass will be available within ten miles, providing both a new market for local farmers and industrial jobs.

Once the demonstration plant validates the technology, Verenium plans to go to full-scale production, building a plant that will produce 30 million gallons of fuel a year. Though initial plant construction costs will still be much higher than those for a corn ethanol plant, says the company's vice president for public affairs, John Howe, at that scale the company expects to achieve a variable production cost of $1.80 a gallon—fully competitive with grain ethanol. It will have to overcome a long history of unmet expectations. Verenium is the product of a 2007 merger between a California enzyme company called Diversa and a Massachusetts cellulosic ethanol company called Celunol. Celunol opened its Jennings pilot plant in 1998, but shut it down for lack of money; three years later the company nearly fell apart and had to be reorganized and recapitalized. Since its inception in 1994, Diversa had by 2007 lost $329.5 million.

Nonetheless, top venture capitalists and the U.S. Department of Energy are investing big money in Verenium's resurrected efforts, believing that both the time and the technology are now ripe. In addition to its new heavyweight CEO, Carlos Riva, Verenium has hired a new vice president of operations, Mark Eichenseer, who as senior resident brewmaster at Anheuser-Busch ran a ten-million-barrel brewery. The company has also licensed its technology to a consortium called BioEthanol Japan, which is operating the world's first plant to make ethanol from wood construction waste.

The first step in the Verenium process tackles the hemicellulose. The biomass is pumped full of water and acid, then exposed to heat and pressure, creating a "steam explosion" that bursts apart the fibers and transforms the hemicellulose into a sweet syrup that resembles molasses. Squeezed out with a screw press, that syrup is put into a tank with a proprietary strain of bacteria capable of fermenting five-carbon sugars.

The damp cake of solids left behind, a kind of sawdust of cellulose and lignin, is washed and the acids neutralized so they do not harm the bacteria downstream. That mash then goes into another vessel, along with bacteria that produce an enzyme which breaks the strands of cellulose into their component sugars (a process called "saccharification," or sweetening) and ferments those sugars into ethanol.

It is this enzymatic step that the scientists at Diversa are now working to radically improve. For more than a decade, Diversa scientists have traveled to the most extreme environments on earth—including deep-sea vents, the soda lakes of Kenya, Siberian volcanoes, and the Amazonian rainforest—to trap "extremophiles," microorganisms that survived by evolving to endure boiling temperatures, crushing pressure, and high acidity and by learning to digest almost anything. The guts of termites and wood-boring beetles are Diversa's current favorite environment for "bioprospecting," says principal scientist Grace DeSantis; full of microorganisms that can ferment wood into methane and hydrogen, they convert 95 percent of what they eat into energy within twenty-four hours.

By now, millions of microbial genomes have been catalogued in Diversa's library. From those genomes, billions of enzymes are mined and screened by robots built by the company, including a "gigamatrix," which can assay a million enzymes at one time. Promising candidates are then optimized with their own patented "directed evolution," which involves trying millions of amino acid permutations and seeing how the mutated enzymes perform.

In their sunlit labs atop a palm-lined hilltop in San Diego, Diversa's scientists tailor enzyme cocktails to use almost any feedstock and spit out many fuels. The company recently signed an agreement to develop wood-to-ethanol processes for New Zealand, and are partners with BP and DuPont on their efforts to make biobutanol. "Diversa is this amazing science machine," says Verenium CEO Riva. "They can do evolution according to a schedule and within financial constraints."

Verenium is also working on its fermentation organisms, evolving them to grow on lower-cost nutrients and to tolerate higher concentrations of ethanol before they die. The final step is distilling the "beer," as it is called, into something more like vodka. That energy-intensive step—which in conventional production of corn ethanol is typically fueled by natural gas or coal, adding to carbon dioxide emissions—is powered by burning the lignin solids to make steam.

Like many of these low-carbon entrepreneurs, Riva seems to be having a very good time. "I did many years of power investment, building global utilities, structuring complicated deals, doing the first independent power projects in the U.S., Mexico, Egypt. Some people tell me I'm doing this as atonement for that work. But it's really because it's the most fascinating thing I've ever worked on. For those electricity-generating projects we always had GE or Westinghouse or ABB to supply the technology. Here we're trying to do that scale but inventing the technology ourselves. And there's potential for innovation across the whole value chain. John Deere is now working on harvesters to chip the feedstocks in the field, which will lower our costs. It's an order of magnitude greater in difficulty, and also an order of magnitude greater in satisfaction. Not a day goes by that I don't learn something new, confront a new challenge.

"We're transforming an economy grown on cheap fossil hydrocarbons into an economy run on cheap carbohydrates that we grow. We know we're making a drop in the bucket, but it's also a revolution."

The biggest hurdle confronting Verenium is by now a familiar one: the need for large-scale financing. "We're doing the 1.4-million-gallon plant to prove to lenders that we can produce at a price competitive with grain ethanol," says Riva, who as CEO at InterGen raised $9 billion of project financing to construct and operate power projects worldwide. "It costs hundreds of millions to build a plant, which means we can't propagate this stuff without Wall Street. We need to encourage the broadest array of technologies. So the key question is, How to loosen billions of dollars in debt financing?"

John Howe believes the answer lies in federal policy. "A carbon cap would monetize the value of strategies that avoid or reduce emissions," he says. "Once such a driver is in place, there will be a much stronger financial incentive to invest in cellulosic relative to grain ethanol, let alone conventional oil production/gasoline refining." Though both he and Riva see the need for short-term bridging policies, including federal loan guarantees and a production tax credit, those will serve, in Howe's view, "as an approximation or brute-force substitute for the carbon cap, which is a more logical, elegant, market-based approach."

IF VERENIUM AND ITS GROWING NUMBER of cellulosic competitors succeed, a whole new array of feedstocks will become possible sources for liquid fuels. By 2009, POET, a South Dakota–based company that is a market leader in corn ethanol, will be able to add the hulls, husks, cobs, stalks, and leaves of the corn plants into its ethanol plant in Emmetsburg, Iowa. Project LIBERTY, as the company calls it, will increase capacity from 50 million gallons a year to 125 million gallons, and increase yields by 27 percent (up to about 500 gallons from each acre, still just a quarter of Verenium's projected yields).

Other companies are investigating the use of fast-growing poplars and willows, a genus of Asian perennial grasses called *Miscanthus* that grows up to thirteen feet high, poultry litter, and the wood waste produced by the logging, lumber, and paper industries. A number of particularly promising candidates may be found among the seventy or so species of shrub in the genus *Jatropham*, which is like the cockroach of the plant world, able to thrive on the most degraded soils, in deserts, rock piles, and trash dumps, requiring little water but living up to forty-five years and producing an inedible oil with an energy content similar to palm oil. A number of growers in India are beginning to cultivate jatropha, and BP is investing $80 million in a joint venture with a British start-up called D1 Oils to convert it to biodiesel.

In the United States, researchers are focused on the native perennial grasses that once extended across the Great Plains. For thousands of years those grasses wove their roots through the soils, creating a dense mat extending ten feet deep. They built those root systems out of carbon dioxide drawn from the atmosphere, storing about a thousand pounds of carbon per year in each acre of soil while improving soil structure, water infiltration, and fertility. The soils those grasses created were so rich, in fact, that when the first sodbusters arrived, their oxen could scarcely get a plow through.

Just 1 percent of that original tall-grass prairie, and just half the carbon originally stored in prairie soils, now remains. But David Tilman, a University of Minnesota ecologist, has for several decades been investigating the possibility of restoring the American prairie, both to take carbon back from the atmosphere into the soil and to produce biomass for energy. These native grasses are in many ways an ideal fuel crop: they are perennials and can be harvested several times a year for a decade without reseeding—in fact, they require little attention but harvesting, which does the necessary work of periodically razing the grass, work once accomplished by wildfires and herds of bison. Reduced energy input means more net energy output, and much greater reductions in carbon emissions. Tilman calculates that biofuels made from mixed prairie grasses could actually be carbon negative, putting more carbon into the soil and roots than would be released during biofuel production and combustion. Because the grasses can be grown on agriculturally degraded lands, they would not displace food production. And rather than diminishing biodiversity, they would enhance it. A move away from fertilizer-intensive agriculture would reduce the nitrogen runoff that now contributes to the creation of immense "dead zones" in the nation's estuaries. And the restoration of a diverse mix of native grasses—big bluestem, sundial lupine, rigid goldenrod, tall blazing star—would have a happy side effect, providing habitat for imperiled birds like the grasshopper sparrow, meadowlark, dickcissel, sedge wren, and other grassland species.

Converting such diverse biomass into fuel, however, will be enormously challenging. Enzymes are specific to particular feedstocks; with some three hundred different plant species per acre in these native grasslands, no one is even close to developing the kind of microbial consortium, or superbug, required to digest the whole mix—one reason why some companies are turning to nonbiological approaches.

Advances in genomics may help in choosing and developing the best feedstocks. Scientists are sequencing the genomes of crops like eucalyptus and foxtail millet (a genetically simpler relative of switchgrass) to enhance selection of natural strains, engineer improvements, and tailor conversion processes; they are also sequencing "biomass degraders" like button mushrooms, which feed off other plants by breaking down their cell walls into sugars. Grain ethanol got a head start by piggybacking on decades of research in the food industry, and the same sophisticated analytics are now being applied to cellulosic crops. Department of Energy chemist Emily Smith, for instance, is exploring the potential of spectrometry—measuring the interaction between light and other kinds of radiation and matter—to determine the ideal time to harvest crops (for instance, when their lignin content is lowest). "Just as vintners monitor and test the sugar content of their grapes in the field, biofuel producers could potentially use this technology to determine if their crop was at optimal development for conversion to ethanol," she explains.

A FEW COMPANIES ARE EXPERIMENTING with genetically engineering plants to produce the enzymes needed to break down their own cellulose. To make sure the plants do not eat themselves alive in the field, they introduce enzymes that become active only at high temperatures (for instance, those produced by bacteria that evolved to survive in hot springs). Another Cambridge-based company, Agrivida, has developed "Greengenes," enzymes that are expressed within the plant but have no activity at ambient conditions in the field: the enzymes are kicked into action by high temperatures within the refinery, where they convert the starch and cellulose into sugars. Stirring just a few bushels of such crops in with other feedstocks could be a much cheaper way of adding enzymes, says Verenium CEO Riva. On the other hand, taking the enzymes into the field raises concerns about transgenic plants, including the potential for cross-fertilization with other plants, dispersing the modified genes beyond their intended use, and the compromise of other kinds of fitness. Plants redesigned for a new purpose may become weaker in their ability to resist disease, for instance, or to remain standing in the face of high winds and other stresses.

Several companies are skipping enzymes and fermentation organisms entirely, relying instead on conversion technologies using catalysts, heat, and pressure. The most common, Fischer-Tropsch, was invented in Germany in the 1920s to turn coal to liquid fuels—a process that is highly controversial today because of the high amounts of carbon dioxide released. But low-carbon biomass, too, can be liquefied using Fischer-Tropsch. It is first gasified by heating it in an oxygen-starved environment; the resulting "syngas" is then reacted over a catalyst to make liquid hydrocarbons. The technology is expensive, in both capital investment and operating costs, but it has two key advantages: facilities can be built today and can use a wide range of biomass inputs. CHOREN Industries, a German company, is building its first commercial plant in Freiberg, Germany, to produce diesel fuel from various agricultural and wood wastes, consuming seventy thousand tons of biomass annually and producing 4.5 million gallons of fuel; CHOREN's American division plans to build a biodiesel plant in the southeastern United States. Eric Larson, a research engineer at the Princeton Environmental Institute, estimates a production cost competitive with $70-a-barrel oil.

A number of companies are now trying to lower those costs. Range Fuels of Broomfield, Colorado, for instance, broke ground in late 2007 for a commercial plant that will turn syngas from biomass into ethanol rather than hydrocarbons, which not only will reduce costs but also will secure for them the 51-cents-per-gallon ethanol subsidy. The company plans to put small, modular units close to cheap biomass sources to cut transportation costs; its first such facility will use "pine scraps" in Soperton, Georgia. (Wood has many advantages as a feedstock: it does not rot easily, it is available year-round, and it is much denser than light grasses, making it easier to move around.)

Though thermochemical systems have the advantage of being able to handle a range of feedstocks without having to remix the enzymes and organisms for each, the biocentric companies remain unconvinced. "It takes billions to build a Fischer-Tropsch plant," says Jack Newman of Amyris. "You have to amortize those costs over the amount of energy the plant will ultimately make, which turns out to be too expensive. You're in effect simulating the forces that made biomass into petroleum in the first place, that intense pressure and heat. But the original forces were tectonic: the weight of mountains plus the core heat of earth. We can just retrofit an ethanol plant, which costs orders of magnitude less."

John Howe of Verenium concedes that "the approach we're taking is costlier right now, no doubt, but there's a lot more room for improvement given that F-T [Fischer-Tropsch] is pretty mature while industrial biotech is still in its early stages of development." He also believes that fuels made with enzymes will ultimately have a much better net energy-and-carbon balance. By definition, enzymes lower the energy required for a chemical reaction to take place. That's their function in organisms, and "that's their magic," says Howe, "in making biofuels."

THERE MAY BE BIOFUELS MAGIC of a very different sort in a process far removed from Verenium's experiments with enzymes. To see one of the most unlikely, and potentially most transformative, biomass-to-fuel experiments, you could visit Phoenix, Arizona. The trip itself provides a quick review of the current strategies for producing energy, and of the mark they leave on the land. Flying in from the north, over the Four Corners region where Utah, Arizona, Colorado, and New Mexico meet, you look down on mile after mile of broad desert mesas, each one's top scraped flat and bare and all linked together by a web of rutted and barren roads. It might take a minute to understand what you are seeing: coal-bed methane rigs atop each denuded mesa, furiously pumping up the natural gas that will soon be burned to make electricity. The seemingly endless sprawl of Phoenix, lights blazing, air-conditioners churning, provides a further review—of where that electricity goes.

But the most surreal part of the journey begins thirty miles west of the city, where brand-new housing developments and emerald-green golf courses give way to saguaro cactus and dusty mesquite. Here, in this desiccated landscape, is one of the greatest electricity-generating hubs in the world. First to rise up on the horizon is the nation's second-biggest power plant, the Palo Verde nuclear facility, each of its three 1,270-megawatt reactors marked by a hulking cooling tower. A bit farther on, Duke Energy's 500-megawatt natural gas plant emerges, then Sempra Energy's 1,250-megawatt plant and another 1,000-megawatt facility operated by the town of Gila Bend. Altogether, more than 9,000 megawatts of electricity is generated in this bleak reach of the Sonoran Desert. Overhead, massive towers strung with 500-kilovolt lines carry a big chunk of that electricity straight to Southern California.

At first glance, the four smokestacks of the Redhawk power plant, the 1,000-megawatt natural gas–fired facility run by Arizona Public Service (APS), seem indistinguishable from the rest. Except that there is a big greenhouse at their feet, and a pipe running from the top of one smokestack into that greenhouse. Rather than venting the carbon dioxide–rich stack gases into the atmosphere, the power plant feeds those gases to a growing family of well-loved creatures. They could be tulips: at the Shell Pernis Refinery outside Rotterdam, engineers use waste carbon dioxide as fertilizer for flower farms, which "adds new meaning," as the *Guardian* quipped, "to the term 'greenhouse gas.'"*

Here, at Redhawk, a far less lovely crop is grown: algae. Beyond a few manufacturers of cosmetics and nutritional supplements, who value them for their high levels of beta-carotene and omega oils, few recognize algae as an object of desire worthy of such hothouse care. Even at Redhawk, facility engineers invest enormous time and ingenuity killing the hundreds of species of algae that invade the cooling towers; because they adapt so quickly, the engineers must regularly cook up a new toxic brew.

Under the right circumstances, however, these microscopic single-cell creatures turn out to be a dream feedstock for making liquid fuels. They are the fastest-growing plants on earth—doubling their mass in a few hours' time. They are the most adaptable, thriving not only in cooling towers but also in sewage, boiling water, ice, Antarctica, and the Dead Sea. They are the richest in high-energy oils ideal for making biodiesel—producing thirty times more vegetable oil per acre than sunflowers or rapeseed—and are rich in carbohydrates that can become ethanol and proteins for animal feed. They filter many air pollutants, neutralizing acids and splitting nitrogen oxides—precursors to smog—into harmless nitrogen and oxygen. Most important, they are the world's most efficient converters of carbon dioxide to oxygen and biomass. Algae have few higher-order functions: They don't need to leaf, flower, produce seeds, or bear fruit. All they do is consume carbon dioxide and divide. "It is photosynthetic life reduced to its essence," says Ray Hobbs, who runs APS's Future Fuels program.

It was their appetite for carbon dioxide that first caught the attention of Isaac Berzin, the chemical engineer who cofounded

---

* *Guardian*, August 12, 2006.

GreenFuel Technologies and began this experiment at APS. A graduate of Ben Gurion University in southern Israel, Berzin was working on a postdoctoral degree at MIT in the lab of Professor Robert S. Langer, a biomedical engineer with more than five hundred patents who was presented the National Medal of Science in 2007. As part of work the lab was doing for the International Space Station, Berzin was growing algae inside a small bioreactor and began thinking about new uses for algae on earth.

While working on his terrestrial system, Berzin was visited in his basement laboratory by a friend, who brought along a friend of his own—a Russian immigrant named Leonard Blavatnik. Then number eighty-seven on *Forbes* magazine's list of the four hundred richest people (by 2007 he had moved up to the forty-fifth place and was worth $7.2 billion), Blavatnik had made his first fortune acquiring stakes in newly privatized Russian companies, including the Tyumen Oil Company, which he later merged with British Petroleum into Russia's second-largest oil company. After seeing what Berzin was doing, Blavatnik offered to write a check on the spot for $100,000. "I'm sorry I can't take your money," Berzin said. "I need at least a million to get this thing from the basement to the roof." "Well," said Blavatnik, "that will take a little longer." After completing due diligence, he returned to Berzin. "Isaac, you asked for $1 million but after looking at all you have to do, I think you'd better take $2 million." So GreenFuel was born. Blavatnik's Access Industries has continued to support the company, along with other venture investors.

Berzin has an unusually affectionate relationship with his algae. "They're not pond scum. They're the sweetest creatures," he told a reporter at National Public Radio in January 2006. He keeps a video of the microscopic creatures on his laptop. "Belly dancing around, they have a little mustache. They touch each other with the mustaches. . . . My kids ask me, 'Oh Daddy, it's so cute. It's like pets. So, what do you do with them in the end?' I say, 'Uh-oh . . . I burn them.'"

In 2004 GreenFuel was ready for it first real-world test of the algae's ability to eat stack gases. On the roof of MIT's 20-megawatt cogeneration plant, Berzin filled 30 nine-foot-high triangles of clear pipe with a kind of algae soup. The experiment proved a great success: Bubbling the plant's flue gases through the mixture removed 82 percent of the carbon dioxide on sunny days and 50 percent when clouds were overhead (photosynthesis, like solar cells, works only when the sun is shining). Night and day, it cut nitrogen oxide emissions by 85 percent. So GreenFuel took the next jump: in 2005 it built its Emissions-to-Biofuel (E2B) algae bioreactor at APS's Redhawk plant.

The first challenge was to find the right algae; most of Green-Fuel's patents, in fact, are for species selection and adaptation. "We look all over the world for nature's finest," says Cary Bullock, GreenFuel's vice president for business development. To protect local ecosystems, the company relies on indigenous varieties or saltwater species (if marine algae were to escape into freshwater, the osmotic pressure would crush them to death, preventing any damage from the alien invasion). They then adapt the algae by gradually shifting their living conditions—the water supply, the chemical makeup of the gas—to match those the algae will encounter inside the farm. The goal is to have algae that reproduce so prolifically—the "local hero," APS's Ray Hobbs calls it—that they crowd out any zooplankton or other algae that might try to move in. A few rival companies are working to genetically engineer a race of super-algae, bigger and fatter and even quicker to reproduce, but Hobbs thinks that is a bad idea. "They want big fat guys they can fish out with a net. But given the dangers of unleashing a GMO [genetically modified organism] that adaptable and prolific, they'll delay themselves fifteen years getting through the regulatory process.

And with twenty thousand species to choose from, why would you need to engineer a new one?"

Arizona, GreenFuel discovered, is great for algae, not least because of the nitrates polluting the Phoenix wastewater, which is used both in the greenhouse and to cool the power plant. For ecosystems, the nitrates are a problem, causing algal blooms that deplete oxygen and create vast dead zones in estuaries and bays. But the same nitrates are great for bioreactors, where an algal bloom is precisely what is desired. Hobbs also wanted to use the nitrogen in the stack gases as fertilizer, skipping the costly pollution-control technology used to remove the nitrogen oxides and instead running the pipe directly to the algae, but regulators wouldn't allow it. "We see algae as agriculture; they see it as power-plant operations. So we had to spend research money on duct work and cranes, and then buy and add nitrogen fertilizers."

The next challenge, which nearly sank the company in 2007 and could still bring it down, is how to configure the bioreactor to allow the algae to get just the right amount of light. Too little is fatal, so either algae on the bottom must be stirred to the top or light must be brought inside the growing thicket. Lots of experiments have gone awry. In Japan, already a commercial producer of algae for supplements, researchers tried lacing fiber-optic cables throughout the tanks to bring light into the depths. The experiment failed: the algae piled onto the tips of the cables and blocked all the light.

Too much light also spells the algae's doom. On a typical 100°F August morning, the sunlight is as intense here as anywhere on the planet. "This much light can burn them to death," says Hobbs. "Algae need rest, like all of us. It has to take in the photon and then rest."

For GreenFuel's first pilot, from August 2005 through November 2006, Berzin solved the problem of "photomodulation" by building giant test tubes lined up in a rack; at the base of each, a spigot could be opened every couple of days to harvest a fraction of the slurry. So how did they mix things up and keep algae from growing on the surface and blocking all the light? "Yeah," says Hobbs, "how about that?" (He can't say more, he says, without revealing protected intellectual property.)

He can say even less about the most recent experiment, called "the 3D matrix system," except to describe its goal: to achieve the highest possible growth per square meter. Outside the greenhouse, the racks still stand where the "seeding culture" was started inside hanging plastic tubes, like giant baggies, which filled with a dark green slime. The algae were then moved inside the secret recesses of the greenhouse itself.

The potential yields from algae dwarf those of any other biofuel crop. Some seventeen hundred U.S. power plants, GreenFuel estimates, have enough land (most of which nobody else wants) for both algae greenhouses and on-site refineries, plus enough waste heat to power both. Unlike soybeans and corn, which can be harvested just once or twice a year, algae multiply so fast they can be harvested daily ("like milking a cow," Berzin says). While an acre of soybeans yields about 60 gallons of biofuels and oil palms about 600 gallons, an acre of algae could yield 5,000 gallons of biodiesel and ethanol a year. At present, only 20 percent of U.S. petroleum consumption is in the form of diesel. But if half of all U.S. cars ran on diesel, as they do in Europe, replacing all of it with soy diesel would require 1.5 billion acres of fertile land, three times the total cropland in the country. Algae could do it in 47 million acres, on land not suited for agriculture. And though they require huge amounts of water, they can tolerate wastewater—and clean power plant emissions along the way.

The potential scale of that carbon cleanup is also immense. In their first configuration, says Hobbs, the algae at Redhawk absorbed 150 tons of carbon dioxide per acre per year. In the newest version, says Berzin, the algae are adding 100 grams of weight per square meter per day, which means they are absorbing 200 grams of carbon dioxide; at scale, that would take up about a ton of carbon per acre per day. At that rate, says Hobbs, algae grown on 1,000 acres of land available at Redhawk would absorb half the carbon dioxide the plant emitted in 2005 and then could be converted to fuel (a coal-fired plant of similar size would require nearly 2,000 acres—a lot of real estate in some regions of the country). If GreenFuel can get to a competitive price for those fuels, he adds, algae farmers will want as much carbon dioxide as they can get their hands on. Hobbs gestures at the landscape around him: empty except for roads and the seemingly endless colony of power plants. "Two thousand– or ten thousand–acre greenhouses; they don't really intimidate me." When the fuel is burned, the carbon will be released to the atmosphere, but its energy will have been harvested twice, and it will have displaced an equivalent amount of fossil fuels.

"Carbon dioxide is a big risk hanging over the head of utilities," says Berzin. "After Al Gore, they feel like everyone sees them as villains. But they need to provide more electricity, because we all keep adding demand. So what do they build? If they build coal, in five years when they have to pay to clean up their carbon they'll look like idiots. If they build natural gas, which is expensive, they'll look like idiots now. But what if they clean up their carbon dioxide, and for every ton of carbon dioxide they get half a ton of algae worth $300 in biofuels? Well, now they're American heroes. If they clean up 20 percent of the carbon dioxide effluent from every power plant in the country, they'll create enough fuel to replace 20 percent of imported oil."

In mid-2007 everything was going swimmingly. GreenFuel had run successful field tests on a small exhaust gas stream from NRG's Big Cajun II 1,500-megawatt coal-fired plant in Louisiana and the Sunflower coal plant in Kansas. (The algae "really voted for Kansas," says Berzin. "And the land: you want 1,000 acres? 10,000 acres? No problem.") In Arizona GreenFuel was setting world records for algae productivity, a fact confirmed in an independent 2007 report by Otto Pulz, president of the European Society of Microalgal Biotechnology. At all three plants the company had successfully produced transportation-grade fuels, and Hobbs was running several APS vehicles on algae diesel.

And then disaster. First came the news that a South African company that had purchased the bioreactor from the MIT roof, claiming it was launching an algae biofuels company and would soon be buying more GreenFuel technology, had in fact used that pilot reactor to con investors to sink money into a company that didn't exist. By the time GreenFuel realized it was never going to get paid, and that its name had been damaged, the fake South African company was facing charges of police bribery and involvement in a carjacking syndicate.

Worse, at Redhawk, the algae had begun growing faster than they could be harvested, choking off light and nutrients and causing them to die. In July 2007 the company's lead investor, Robert Metcalfe of Polaris Venture Partners, stepped in as CEO, shut down the greenhouse, and laid off half the staff. "Harvest was just one of the issues," says Hobbs. "The underlying problem was that this little start-up was growing really fast and had a communications breakdown between the scientists and the guys on the ground." Jennifer Fonstad, from Draper Fisher Jurvetson, another backer, became chairman of the board; Berzin remained chief technology officer. Metcalfe, who in the 1970s invented the Ethernet for local-area networking and cofounded the billion-dollar networking company 3Com, called the breakdown a "success failure" typical of efforts to commercialize emerging technologies. To keep the company going, he raised more funding. "I keep asking the trillion-dollar question that led to the founding of GreenFuel," he explained. "Why expensively sequester carbon dioxide when it can be profitably recycled?"

Perhaps more than any other cutting-edge green technology, algae arouse intense passions, including intense skepticism; several of the Web sites that track clean technology had great fun with the GreenFuel setback and claimed vindication of their doubts. Many likened it to the collapse a decade ago of the U.S. government's Aquatic Species Program, which beginning in 1978 had conducted algae research for nearly twenty years. Using open ponds rather than closed bioreactors (a strategy now being pursued by several GreenFuel rivals), the federal algae farmers had been overrun by wild species that outcompeted their obese, domesticated cousins. In 1996 the Department of Energy concluded that algae fuels were too expensive to produce at large scale and shut down the program. (Late in 2007, with oil prices topping $90 a barrel, the National Renewable Energy Laboratory announced that the government had changed its mind and would resume research into algae.)

Berzin believes the Energy Department made a fatal mistake in not taking its research to a power plant to make use of wastewater, flue gases, and the waste heat available for drying and refining the algae. "They never had that complement, were never exposed to that potential for integration," he says. Ray Hobbs is more sarcastic. "Oh well, I guess if the government can't do it, then nobody can."

Hobbs also does not acknowledge defeat. "GreenFuel jumped in scale too fast, to a thousand-square-meter greenhouse, and were caught unprepared. But when Isaac [Berzin] regrouped and rescaled to a hundred square meters, it worked like a charm. When you learn why something doesn't work and you fix it, that's a successful experiment, not a failed one. It's like Edison. After 999 filaments he found the one that worked. And he also found all the ones that don't work. If you succeed the first time, you won't really understand why."

The venture capitalists who are backing GreenFuel have given the company one more chance to prove the technology, imposing a discipline Berzin finds bracing. "They don't care whether something is intellectually interesting," he explains. "They want to know if it works. If you're an academic, you're like a medical doctor. If you make a mistake, your patient dies. But if you're a start-up, you're like the pilot of a small plane. If you make a mistake, you die."

Even if GreenFuel crashes, Hobbs is prepared to pick up the wreckage. "If GreenFuel in its current form folds because the VCs [venture capitalists] lose patience, I just see us regrouping. The company might go under, but not the idea. APS is 120 years old. If we have to go this alone, with the utility capability, we will. Their interest is the profits to be made by making fuel and competing with oil; our interest is in a better way to capture carbon." That level of commitment, says Berzin, has made all the difference. "There were many Isaacs before, but there was never a Ray before. APS was the first commercial energy company to open these doors."

By late 2007, GreenFuel had begun deployment of a sixth system Berzin calls a "horizontal thin film," which uses more land but promises to cut capital costs enough to beat oil at $60 a barrel. He is also working on cheaper ways to separate the algae from the water and extract the oils to be made into fuel. He has partnerships with leaders in global "algaculture," including an Israeli company that was among the first to use flue gases, but all those partners grow high-value algae that sells for $5,000 a ton or more. For fuel, Berzin needs to cut that price by an order of magnitude. That means he cannot afford to pressurize the carbon dioxide to bubble it through the algae, or to use a centrifuge to separate the algae from the water. But "after $26 million and six years of growing gray hairs," he says, he has invented most of the operational units he requires, albeit quietly. "If you're a scientist who makes a breakthrough you want to run naked through the streets shouting. If you're in business, you keep your mouth shut and run to your patent lawyer instead."

HOBBS SEES THE EXTRACTION PROBLEM as relatively trivial. "You add a little methanol and a catalyst, you get glycerins and methylesters. You hydrogenate the methylesters, you get diesel fuel. We'll partner with GE Water or 3M Filtration Systems. We're not three guys in a lab. We're going to do this like we do things. Then we'll ship the leftovers to shrimp farms for feed, or ethanol plants. Both have already come knocking."

The multiple iterations, explains Berzin, are not a matter of discarding old experiments for new, but rather the development of a portfolio of options to address differing circumstances. In Europe, where land is expensive, power plants could use the more expensive "matrix" to maximize productivity per acre. Where land is abundant and cheap, as at most U.S. power plants, they could use the thin-film technology to maximize returns.

GreenFuel will now tackle one of the largest sources of global warming pollution in the West: the Four Corners Power Plant west of Farmington, New Mexico, owned by APS and five other utilities. On a return flight from the moon, *Apollo* astronauts reported seeing two human structures from space—the Great Wall of China and the plume from this 2,000-megawatt coal-burning plant. The Environmental Integrity Project ranks it the very worst among power plants for total nitrogen oxide emissions and eighteenth for carbon dioxide emissions. Before feeding the flue gases to the algae, GreenFuel may also need to do extra scrubbing of sulfur dioxide, which acidifies in water and might corrode the duct work, and to select algae that do not absorb heavy metals like mercury. "Algae enjoy a nice coal flue gas," says Hobbs. "It has lots of nutrition, and they have an amazing capacity to handle contamination. But it might be too much, and we don't want the dangerous stuff."

The Four Corners algae project is a partnership among seven

utilities in four states: APS, El Paso, the Los Angeles Department of Water and Power, Southern California Edison, Tucson Electric, Salt River, and Public Service of New Mexico. The land is being leased from the Navajo, who have a hundred thousand acres under irrigation but have not yet found the ideal crop to grow. The carbon cap in California, which extends to out-of-state power suppliers, has pushed the project forward, says Hobbs. Arizona itself is part of the Western Climate Initiative, along with California, New Mexico, Oregon, Utah, Washington, and the Canadian provinces of British Columbia and Manitoba; all have agreed to cut carbon emissions across the region 15 percent below 2005 levels by 2020.

The effort to make fuels from algae fed on stack gases is not Hobbs's first foray into alternative transportation. Almost twenty years ago, working as an engineer on APS power plants, he persuaded his boss to let him build an electric race car. His team converted a Honda, and when in 1991 it won the inaugural Solar and Electric 500 at the Phoenix International Raceway, it made the foldout in *Popular Science*. "That set a fire in Detroit. The car companies were bitterly resistant. They feared that GE or Siemens would move in on them with their huge financial muscle. They didn't want to have to remake their manufacturing lines and retrain their dealers. And mostly they didn't like getting told what to do.

"GM was especially hostile toward us. But everyone told me they were the key to moving the industry, so the next car we converted was a Saturn. Motorola had asked if they could develop our electric drive. Those first brainstorming sessions with them just blew me away. I wasn't used to a competitive industry; at utilities we just kind of hmm-hmmm along."

Again, the electric Saturn dusted the rest of the field, accelerating from 0 to 60 miles an hour in 4.7 seconds. In 1994 the team beat that record again, converting an Indy car and hitting 60 in 3.2 seconds. "That one was scary to drive—like a silent rocket," says Hobbs. "It had maximum torque at zero rpm, so it cornered amazingly. The first time it came past us at 120 miles per hour, we heard a weird chirping noise. We finally realized it was the tires on the pavement, which no one had ever heard before because of the engine roar."

Lately, Hobbs has stayed away from the racetrack. "We got way ahead of the car companies, which was never our intent. Rockefeller helped develop the internal combustion engine to increase demand for oil. We just wanted to make a market for our off-peak assets."

In fact, Hobbs's motivation for his experiments seems much deeper. "We're able to live the way we do now because the decisionmakers in our parents' and grandparents' generations thought about us and built the infrastructure we would need," he says. He likens the present generation to the grasshopper in Aesop's fable— singing all summer while the ant stored up food, then starving when winter came. "When did we become grasshoppers?" he asks. "Every human takes care of the generations behind them."

Hobbs's optimism occasionally deserts him. "If we all fail, and we might, the rich and strong will have the last safe places on earth. And the billions of poor people will either figure out a way to take those places back, or they will die."

But against that despair Hobbs marshals a deep reverence for nature. He is awed by algae. "You are looking at the origins of life, an organism that has survived for three and a half billion years and created the conditions for other life to emerge. They are the root of the food chain. And so elegant. Single-celled algae can crack water with a photon into hydrogen and oxygen, then metabolize that hydrogen with carbon dioxide to sugar. We can't do that. We can't even fully understand it. You just have to be mesmerized and humbled."

Like Metcalfe, Hobbs thinks the idea of sequestering carbon underground is both wasteful and risky. "Carbon dioxide is dangerous in large quantities," he explains. "We spend enormous amounts of money to dig holes to get carbon out of the ground, and the best fix we can come up with is to dig more holes and put it back in? That's something your dog might think to do.

"Given all we invest to get carbon in the first place, shouldn't we think of ways to use it again? The tomatoes in my sandwich come from greenhouses where they burn natural gas just to make carbon dioxide to feed the tomatoes. We have a valuable raw material coming out of that smokestack, and we're just throwing it away."

Hobbs spends a lot of time thinking about the carbon cycle and how it has operated since the emergence of life on earth: carbon moves from the atmosphere into living organisms and the oceans, then into sediments and rock, and only very slowly out of the rock and back into the atmosphere. That cycle was in balance until humans started digging up and burning deeply buried reservoirs of ancient carbon; without our thirst for those fossil fuels—the fossilized remains of plants and animals that predated the dinosaurs— that ancient carbon would have cycled back into the atmosphere very slowly, over many eons.

"The answers are right in front of us," Hobbs argues. "Algae is nature's building block—it's how carbon was managed on the planet for hundreds of millions of years. Can't we seize that example? By digging up fossil fuels that were stored over millions of years, humans accelerated the front end of the carbon cycle. So we're accelerating the back end, the uptake part of the cycle. From the time we burn the carbon in the power plant to the time we get oil out of the algae is a matter of a few hours. We get that carbon back *in a few hours*. And we keep thinking about just how many times we could get that same carbon to play. After a while, you could lose the coal entirely. You could grow the algae, burn the algae in the power plant, capture that carbon dioxide and pump it back into the greenhouse, and grow more algae. Dried algae has as many BTUs per pound as coal. You'd have a completely renewable power plant.

"You have to gather it all up in your arms, think of it as a whole system. We throw away heat. We need to use it. We throw away flue gas; we need to use that too. But these ideas will only work when carbon dioxide has monetary value. Without a price on carbon dioxide, the economics will always pressure you to burn fossil fuels. No one knows how to give that up, it's our economic universe.

"And if you don't get long-term policies in place, the political momentum we've got going right now will just dump out the other end. If we get this, the oil companies will go nuts, because it's distributed, which means they can't control the supply. They'll send their lawyers and lobbyists in, just like they did with the electric car. The political allies we have now won't be there long enough. I lived through the zero-emissions mandate—and saw the government fold under pressure.

"You need a way for people to make money. Forget wisdom. We have to play off greed. A carbon cap will be the turning point. You can't stack the deck against these scientists and innovators. Society has to stand with them. And utilities are nothing more than instruments of public policy.

"I have a devout belief in God, and I believe God put carbon in the bank for us to use at the right time, but that we have to rise to the challenge to manage it."

CHAPTER 6

Ocean Energy

Few places overflow with life as generously as the Makah
Bay, at the northwest tip of the continental United States
near Olympic National Park. Soaked by as much as two
hundred inches of rain each year, the temperate rainforests on this
remote peninsula are so fertile and lush that even the cool, moist
air seems green and alive. Immense, old-growth cedar trees climb
down the mountains to the sea, laden with mosses that sometimes
grow heavy enough to break branches; lichens, mushrooms, and
ferns colonize every stone and fallen log; pink-orange bursts of
salmonberries flash among the leaves. On the beach, the dock, and
the pilings, bald eagles loiter like mere seagulls, a dozen at a time.

The waters off that beach hold more life still. Where enormous
waves regularly pound this shore, wrecking hundreds of ships over
the centuries and carving deep caves into the black cliffs, a sea otter
has found a quiet corner and lolls on its back, snacking. In the shal-
lows, thick with algae, a gray whale feeds, the arc of its speckled
back and spouting blowhole marking its languid journey. Beneath
the surface, where the continental shelf meets three offshore can-
yons, deep, nutrient-rich Pacific waters mix in the "Big Eddy" with
the sediment-laden flow of the Fraser River, tumbling down from
the Canadian Rockies, creating one of the most productive and

diverse marine ecosystems in the world. An unmatched variety of seaweeds, algae, and invertebrates live here, as well as glass sponge and black and gorgonian corals. Among the foraging seabirds are the threatened marbeled murrelet, rhinoceros auklet, and tufted puffin. Elephant seals, harbor seals, and Steller sea lions mingle with the fifteen species of whales that migrate through, including humpbacks, orcas, and fins.

For several thousand years, this marine abundance has supported the Makah tribe, several hundred of whom live in the little town of Neah Bay, one of the very few towns scattered along this isolated coastline. Paddling as far as a hundred miles from shore in canoes hewn from single logs of cedar, dozens of generations of Makah have fished and hunted seals and whales for their meat and skins. When, in 1855, the tribe signed the Treaty of Neah Bay with the governor of the Washington territory, tribal leaders gave up three hundred thousand acres of the most valuable timberland in America to preserve the "right of taking fish and of whaling or sealing," which they considered then and now to be the core of their culture.

Though its population and culture declined precipitously during the century that followed, several events over the past thirty years have served to revitalize the tribe and its relation with the sea. In a 1974 ruling considered among the century's most important on Native American rights, U.S. District Judge George Boldt found that the Makah treaty guarantees them half the salmon and steelhead harvest in their "usual and accustomed" fishing grounds, covering five hundred square miles, and that the tribe had the primary authority to manage those fisheries. That decision, largely upheld by the Supreme Court in 1979, established the Makah as comanagers of the fisheries with the state and federal governments.

In that same decade, archaeologists unearthed the pre–Columbian Makah village of Ozette, a few miles south of Neah Bay, which like much of Pompeii had been perfectly preserved when it was buried—in this case by a mudslide. That discovery of some fifty-five thousand artifacts further reinvigorated the Makah's sense of ancestral identity. After the gray whale was deemed sufficiently recovered to be removed from the endangered species list in 1994, the tribe determined to revive what it considers its most fundamental tradition. Eight men began the rigorous physical and spiritual training necessary to undertake the first whale hunt—in the traditional manner, with harpoon and canoe—in seventy years. Sustained protests against the 1999 hunt, some of them violent, made national news.

Today, the sea remains the center of Makah identity. Their tribal totem is a thunderbird holding a whale in its talons; in the summer evenings, war-canoe racers practice in the harbor. Fishing remains the tribe's main source of income. In the spring, the town is flush, with thirty-five small boats landing mountains of halibut and salmon. As the days shorten, and unemployment climbs toward 70 percent, things get harder. A third of the town lives below the poverty line, many in dilapidated trailer homes; in the 2000 census, the median family income was $21,625 a year. As winter storms roll in, bringing hundred-mile-an-hour winds and fearsome blizzards, power lines go down and the heat and lights often go out. In 2006–2007 the village spent twenty-five days without power, in darkness and freezing cold.

IN 2001 THE MAKAH TRIBAL COUNCIL found a new way the sea could sustain its people. They were visited by an energetic woman with a little start-up company that, she said, could turn the kinetic energy in their world-renowned waves into electricity.

Alla Weinstein emigrated from the Soviet Union in 1974. After twenty-two years at Honeywell Aviation designing navigational controls and installing systems on Russian airplanes, she became involved in wave energy in the fall of 2000, when she met Hans and Gören Fredrikson, sons of the man who had invented the technology she was offering to bring to the Makah. Her favorite expression, delivered with a twinkle, is "It's all about the motion in the ocean." The name of her company, fittingly: AquaEnergy Group.

About three miles out from shore, she told tribal leaders, a cluster of buoys would be moored loosely to the seafloor. Attached to the bottom of each buoy would be an eighty-foot steel cylinder, "an acceleration tube" submerged vertically in the water and open at both ends. In the middle of each cylinder would be a piston and, attached to the piston, two rubber marine hoses, one connecting to the top of the cylinder and the other to the bottom. As the buoy rose and fell on the waves, the piston would lag behind, slowed by the water in the cylinder, causing the hoses to stretch and relax in turn. Like a cow's udder, or like the little Chinese fingertrap that you slip on and can't pull off, each stretched hose would contract like a muscle, pumping water at continuous high pressure (the equivalent of a 650-foot waterfall) into a turbine and generator (preferably a centralized turbine, floating on a barge nearby). A standard submarine cable would then carry the electricity back to shore. The device would be durable, she assured them: hose pumps have no seals or moving parts, and are actually lubricated by saltwater. And in major storms, the slack mooring would allow the buoys to ride tsunami-size waves out at sea, rather than be crushed by them as they broke onshore. Most important, the modular design—with each buoy independently producing up to 250 kilowatts of electricity—would protect the power supply and simplify maintenance: if one failed, the company would not need to maintain it at sea but could simply lift it out and replace it.

Just four buoys, Weinstein told the five members of the tribe's leadership council, would provide a megawatt of electricity, enough for 150 Neah Bay homes. In time, when the pilot plant had proved itself and the company had optimized the technology, it could add sixteen more to meet the village's total electricity demand. The tribe could even make electricity to sell: an 80-megawatt plant, arrayed in a starfish configuration with ten buoys on each arm, would require a quarter of a square mile of the bay and supply half the electricity for the whole Olympic Peninsula.

Weinstein's timing was perfect. The tribe had recently determined to become energy self-sufficient and to diversify its struggling economy, and had begun investigating windmills and the use of wood waste as fuel. How much better to turn to the resource that had always sustained them? In 1994 the tribe had rejected another kind of energy development in its waters: when the U.S. Minerals Management Service began considering reactivating leases for offshore gas and oil development, the Makah supported the creation of the Olympic Coast National Marine Sanctuary within its treaty-protected waters. A 1991 fuel spill from the Japanese fishing vessel *Tenyo Maru* had damaged ecosystems all along the Washington coast, heightening the tribe's concerns about offshore production and oil transport. The Strait of Juan de Fuca, on the village's northern flank, is the second most active waterway in the world; every year, fourteen thousand vessels pass through into Puget Sound to the ports of Seattle, Portland, and Vancouver, many of them carrying oil from the North Slope.

AquaEnergy seemed amenable to an equal partnership. Tribal chairman Ben Johnson, a fisherman who over many decades had gained intimate knowledge—much of it passed down by his father—of the habits of various species and the seasonal patterns of the waves, pored over maps with Weinstein to choose the best spot for the buoys, making sure to skirt prime fishing areas. They discussed terms, with the tribe making clear that eventually they would switch from upland leaser to equity ownership.

Finding a buyer for the energy was easy. The Clallam County

Public Utility District, eager to develop more diverse and stable energy supplies, agreed to purchase the electricity and deliver it through its grid to Neah Bay. If, eventually, the enterprise goes to commercial scale, an upgrade of transmission lines will be required. The feeder lines serving Neah Bay now handle just 14 megawatts, not the 80 megawatts AquaEnergy ultimately hoped to produce.

Securing the necessary permits—some fifteen in all, from state and federal fish and wildlife agencies, the U.S. Army Corps of Engineers, the state historic preservation officer, the Federal Energy Regulatory Commission (FERC), and the marine sanctuary, among others—has been a far more protracted affair. Part of the problem, especially at first, was AquaEnergy's lack of experience with the regulatory process, and its lack of money. Just as Weinstein was founding the company, the dot-com bubble burst and venture capital dried up. "Most of that money," she says "wasn't interested in infrastructure anyway." She found herself in a catch-22: without permits she couldn't get financing, and without financing she couldn't do the studies needed to secure the permits.

She might have given up were it not for a tragedy that fortified her determination to succeed. On December 11, 2001, Weinstein was in Neah Bay for a technical meeting with the tribe, awaiting the arrival of her brother and partner, Yury Avrutin, and company cofounder and chief technologist Bengt-Olov Sjostrom. The two men were in a small plane, scouting potential sites on the California and Oregon coasts along the way. They never arrived. From then on, says Weinstein, a warm and effusive woman, failure ceased to be an option. "Maybe events like that push you to do things you might not otherwise do. Why else would I put myself in this position; so much uncertainty, no income for six years? I need to get it done. Otherwise they lost their lives for nothing." In honor

of their memory, she gave the device a new, superhero name—
AquaBuOY—capitalizing the initials of her lost colleagues.

To build consensus around the project, she began meeting
with commercial and recreational fishermen, local environmental
groups, the marine sanctuary advisory committee, the Surfrider
Foundation, the Northwest Energy Coalition, and any other group
that voiced interest or concern. In 2003 she went to the Clallam
County Fair and collected one thousand signatures from citizens in
support of the project, which she delivered to the state of Washington's congressional delegation.

As FERC considered the license application, the public com-
ment period brought hundreds of responses. An association of Oregon crabbers expressed concern about the possible loss of access
to productive grounds. But its was the sole unambivalent objec-
tion, according to sanctuary superintendent Carol Bernthal; others urged only that the project be managed with the utmost care.
Many, she says, were simply fascinated by the technology. Environmental Defense Fund weighed in, with a letter signed by marine
biologist Rod Fujita and representatives from four other conservation groups voicing concerns about impacts on the marine habitat
and erosion of sanctuary rules, and urging that a full environmental
impact statement (EIS) be required.

Things sped up in 2006, when Finavera Renewables, a Canadian
company with roots in Ireland, purchased AquaEnergy. Finavera
hired a lawyer with permitting experience and brought in SAIC,
an $8 billion engineering firm (its core business is national defense)
with an aquatic environmental team of two hundred marine biolo-
gists, oceanographers, water and sediment quality experts, cultural
resource archaeologists, and hydrodynamics scientists with thirty
years of experience doing environmental assessments and monitoring. In November 2006, Finavera applied for the nation's first pre-

liminary permit for an offshore wave energy plant. In April 2007 it was granted.

As of this writing, one of the hurdles remaining was securing the permit from the marine sanctuary, which has found itself in the difficult position of assessing an industry that has almost no track record and a technology that is not yet fully designed. "We are having to rely on analogy," says Bernthal. "And not knowing, in a place of national significance, has pretty high stakes."

A primary sanctuary mandate is to protect the seafloor, so sanctuary officials are paying particular attention to the anchors that will secure the buoys and transmission cable. If the cable traverses bedrock, where creatures attach and create the three-dimensional habitat structure critical to fisheries, anchors could do serious harm. The sanctuary and tribe have already been burned once. In 1999, the high-flying company Global Crossing buried sixty-six miles of a fiber-optic cable that links the United States with Japan within the sanctuary—carelessly: unmoored stretches of cable ended up obstructing access to the tribal fisheries. Then Global Crossing went bankrupt and left the mess behind; only in 2006 was the cable finally repaired and reinstalled. To insure against a repeat of that experience, the sanctuary will consider bonding this time around.

If FERC's final environmental assessment fails to address all of its concerns, the sanctuary will do its own supplemental assessment. "Our level of scrutiny is high compared to FERC, which normally deals with large-scale hydroelectric projects," says Bernthal. "They may issue a license without final engineering, and speculate on the impacts. We need more than that." If the sanctuary does issue a permit, it will be for only the five-year pilot project, with the assumption that the buoys will then be decommissioned and removed. Because a larger-scale project may have additional impacts, such as disruption of local circulation patterns critical for transport of larvae and nutrients, a decision to pursue a commercial-scale project will require the tribe and the company to start the permitting process again. Throughout the pilot project, the sanctuary will require rigorous monitoring, "so that five years from now we'll be able to say whether in fact these projects are benign." What is most important, says Bernthal, is that "we not greenwash this stuff, but have a real, honest dialogue about the trade-offs."

Finavera's engineers have continued revising the design to address environmental concerns. They have reshaped the AquaBuOY to avoid tempting seabirds or sea lions to perch. Although their hydraulic circuit uses water, not petroleum-based fluids that could hurt marine mammals and sea birds if spilled, they have sealed it so fish cannot be trapped inside. They have found a path for the cable entirely on soft bottom, where large movements of sediment prevent any creatures from taking up residence (the sand moves so quickly, in fact, that several of the devices they laid down to measure currents have already vanished). When told that the mooring chains from the concrete anchors they planned to use might sweep across the floor and hurt the "benthic community"—the crabs, sponges, snails, and other small organisms that live in bottom sediments—they proposed a switch to "vertical launch anchors," recently developed for offshore oil-drilling platforms. The broad, flat anchors, resembling sting rays, are embedded beneath the seafloor, reducing their surface footprint to the thickness of the cables connecting to the buoys, and can also be unlatched and retrieved. Since the design is not yet complete, Finavera will try to persuade the sanctuary to provide a contingent permit, revocable should the company fail to provide a satisfactory solution.

The tribal leadership bristles at mention of the sanctuary's role. "If it weren't for us the sanctuary would not be there," says Johnson, who as chairman is shown the deference awarded the hereditary chief before the treaty abolished the traditional leadership system. "They say we have fishing rights in the sanctuary. But the sanctuary is in our territory. We were here thousands of years before them, and allowed them to come into our waters. The designation document for the sanctuary has a savings clause, protecting our economic interests."

Councilman Micah McCarty sees a parallel with the struggle the tribe faced over the resumption of whaling. "We are an ancient society that still has a living relationship with our ancestral fishing and hunting grounds. By continuing to sustain ourselves from these resources, we keep the breath of our ancestors alive. It has spiritual meaning. But we repeatedly run up against a belief that nature should be viewed without touching it, kept pristine. I understand where that view derives—it comes from people who live in a wholly altered environment, see a devastating human impact, and overcompensate for that devastation. But it winds up disenfranchising the people who depend on the land."

MAKAH BAY IS A KIND OF CRUCIBLE, within which all the tensions around the emerging new-energy economy are distilled. Though here the tensions are heightened by the exceptional value of the ecosystem, the jurisdictional complexities, and the acute need for economic development and reliable energy, none of the new carbon-cutting technologies will be without environmental impact. All will require vigilance and a balancing of competing environmental values. Weinstein argues that warming and acidification of the oceans from business as usual and continued climate change will do far more damage to northwest Pacific ecosystems than her cables and anchors.

Carol Bernthal wonders whether a marine sanctuary, one of only fourteen in the entire country, is the best place to undertake this experiment. She also recognizes that the Makah are tied to their traditional waters and cannot do the project elsewhere, and that the energy that will be displaced comes from immense hydroelectric dams, which—even though they emit no carbon—have destroyed most of the salmon habitat in the Northwest by changing water temperatures, flows, and bottom habitat. The sanctuary is already stretched beyond its resources, and cannot itself assess the impact of this project. But where better to hold this new industry to the highest standards?

One member of the Makah tribe, fortunately, is extraordinarily well equipped to reconcile these many tensions. A graduate of the George Washington University Law School, Robert Martin spent twenty-three years in Washington, D.C., leading an association of tribes managing mineral-rich lands, representing tribes in environmental fights, and, finally, serving three administrations as an independent ombudsman at the U.S. Environmental Protection Agency (EPA). His job was to investigate public complaints against the agency, and he became, in that position, something of an icon: in her 2003 book, *Bushwhacked*, the late Molly Ivins devoted an entire chapter to him. Describing him, aptly, as "a bear of a man" who with his long ponytail and rumpled clothes sometimes "looks like an unmade bed," she told the story of his 1992 victory helping small-town newspaper publisher Marie Flickinger stop the EPA from incinerating 245,000 tons of toxic sludge at a Texas superfund site. Though Flickinger had been petitioning the EPA for five years without success, within a few months after Martin's arrival the agency had sealed the site, mandated that the water be treated, and evacuated a contaminated elementary school and 677 homes nearby. As Ivins put it, he used his first big case "to expand and define the powers of an office that was almost an experiment when he drew his first paycheck."

In 2006 Martin returned to Neah Bay to help his tribe pursue its energy plans and additional opportunities he sees emerging in a carbon-constrained world. In the tribe's forests, for instance, it might do "carbon farming": growing and preserving timber to store carbon, then selling those "offsets" in the carbon market. Martin's environmental credentials remain sound: he is on the board of the Natural Resources Council of America, a group of more than eighty-five environmental groups that includes Ducks Unlimited, the Humane Society of the United States, the Ocean Conservancy, and the Woods Hole Research Center. He sees the permitting issue with the sanctuary and other agencies "not as obstacles, but as necessary conversations."

THE MAKAH PROJECT has a number of important allies. The Clallam County Economic Development Council strongly supports the program, calling support for innovation "the basis of rural development." The council estimates that of the $4 million it will cost to complete the Makah Bay pilot project, half will stay in the community. Given its maritime history, the county has the local suppliers and most of the marine engineering capability needed to assemble, install, and maintain the buoys.

Many in the Northwest share the view that the development of ocean energy offers the best opportunity to rebuild coastal communities and maritime trades that declined along with the fishing, shipbuilding, and logging industries. Not surprisingly, much of the political leadership in the region is working to accelerate that rebuilding. Democratic U.S. Senators Maria Cantwell and Patty Murray of Washington, Ron Wyden of Oregon, and Daniel Akaka of Hawaii and Republican Senator Lisa Murkowski of Alaska are pressing, along with Democratic Representatives Jay Inslee of Washington and Darlene Hoorley of Oregon, for legislation that would provide hundreds of millions of dollars in funding for wave energy research and development, as well as tax credits and loans. In 2006 Oregon state legislators put wave energy at the top of their economic development agenda and passed a generous investment tax credit; for Finavera, that credit will equate to a refund of 25 percent of expenditures.

California has become particularly active on the issue, attracting big companies with money and muscle, thanks in large measure to the strict caps on global warming emissions passed by the state legislature in 2006. PG&E, for instance, has applied for FERC licenses to build two 40-megawatt wave farms by 2010. In 68 square miles off Fort Bragg in Mendocino County and 136 square miles off Humboldt County near Eureka, the utility will spend $3 million testing technologies. In late 2007 it signed a 2-megawatt power purchase agreement with Finavera.

Oregon Iron Works (OIW), a sixty-five-year-old metal fabricator with more than four hundred employees, is already benefiting from this emerging industry. In its sheds in Clackamas, near Portland, the company built the prototype buoy for Finavera and is building a second prototype for a competitor, New Jersey–based Ocean Power Technologies. OIW does a range of prototypes and production runs, many for the U.S. military, including boats, hydroelectric dams, fish screens, containment vessels for nuclear waste, the launch complex for Atlas and Delta rockets, and an electric streetcar. Its experience building for the unique challenges of marine environments—forces and torques far more powerful than those endured by windmills, corrosive saltwater that eats at welds—has proved useful in developing these wave energy devices, says vice president Chandra Brown; OIW knows, for instance, how to create seals with super-close tolerances that remain tight under extreme water pressures.

Although it performed as hoped for two months, the first test AquaBuOY, deployed in August 2007, came to a bad end. Just one day before it was to be removed, it sank to the bottom of Makah Bay. Finavera explained that a few days before it sank, the buoy began taking on water—then its bilge pump failed. The company pronounced the test a success, nonetheless; all through its deployment, the buoy had been supplying valuable data that will be used in developing the next model.

Beyond ensuring that the next generation stays afloat, the fabricators face two additional challenges. First, OIW needs to help Finavera optimize the AquaBuOY's performance. In projecting the output of the four demonstration buoys planned for Makah Bay—1,500 megawatt-hours per year—Finavera assumed that they would produce electricity at just 18 percent of capacity; the goal for optimized production installations is 40 to 50 percent capacity. The second challenge is to cut costs, which for the prototype now come to about 25 cents per kilowatt-hour. "It took wind twenty-five years to get from 30 cents to 7 cents," says Brown. "We plan on a much quicker learning curve." If OIW succeeds, and lands contracts to mass-produce buoys for the two companies, it could quickly become a key player in the wave energy industry. "We've intentionally stayed a small business," says Brown, "but we are now talking about kicking off a new division. We believe the ocean is one of the biggest untapped resources. We also believe that U.S. manufacturing needs to be alive and well. We don't want to be outsourcing; we want to be in-sourcing. Keeping those skills sets, and developing domestic renewable energy sources, are both security issues. We can't depend for our infrastructure on China, and we can't depend for our power on foreign sources."

THOUGH THE FINANCING and development of ocean energy have lagged beind solar power and biofuels, recognition of its potential is growing. A 2005 report from the International Energy Agency notes that energy exists in the oceans in several different forms, the most important of which are marine currents, caused by

variations in salinity and temperature, and waves, which are generated by surface winds.★

Because the best waves are created by the steadiest, strongest winds,
the wave resource is largest on western seaboards between 40 and
60 degrees latitude. Prime areas are found off the densely populated
Pacific Coast from northern California to Alaska, Europe's Atlantic
coasts, western Australia, and the southwest coasts of South America and Africa. (At some specific sites, such as Makah Bay, the shape
of the shoreline and underwater features serve to concentrate the
wave energy, making them especially attractive for development.)
The Electric Power Research Institute (EPRI) estimates that wave
power could eventually meet 10 percent of total U.S. demand; similar estimates are made for the global potential. Advances in technology may expand the regions in which the resource can be used,
as they have for wind. While the early-stage systems will require
good wave structure and big waves, eventually the industry may be
able to harvest energy from Atlantic swells.

THOUGH ITS TOTAL POTENTIAL is likely only a fraction
of solar's, ocean energy does have several distinct advantages. One
is consistency. Waves steadily pound the coasts because the ocean
serves as a big energy storage system. It does not matter where in
the Pacific the wind is blowing: once generated, waves are preserved and can cross the entire ocean without losing much energy.

A second advantage is predictability. Data buoys all over the ocean
provide several days' notice of when waves are going to hit the coast,
allowing grid managers to plan accordingly. Mariners rely on the
National Oceanic and Atmospheric Administration (NOAA) fore-

★ A third potential way to make energy from the ocean would use the temperature
difference between warm surface water and the cold depths much as the Chena,
Alaska, geothermal power plant uses the difference between cold creek water and hot
springs (see chapter 7). It remains unclear whether this "ocean thermal energy conversion" technology will make a major contribution to the future power mix.

casts, Wavewatch III; in 2007, EPRI began a study to refine correlations between deep ocean buoys and ones nearer to shore.

A third advantage is energy density. Waves are a third-order power source: the sun produces winds (by differential heating of the earth); winds in turn transfer that energy to the water, which is eight hundred times denser than air. At each step the energy gets more concentrated. So while the maximum solar energy per square meter is about 1,000 watts and the maximum wind energy is 10,000 watts, the maximum for waves is as high as 100,000 watts, higher still during storms.

As with wind power, the Europeans are ahead of the United States in developing the new hydroelectric technologies, supporting them with subsidies and infrastructure assistance and, most important, setting caps on carbon emissions. For its project in Portugal, Finavera Renewables secured a 1.3-million-euro grant from the European Commission, the executive body of the European Union. Portugal also pays 40 cents (American) per kilowatt-hour for wave energy. That above-market rate—called a "feed-in tariff" since the government is giving rather than taking the tariff—will decline over time. The South West of England Regional Development Agency (SWERDA) is spending $43 million to build a wave hub off the coast of Cornwall: a high-voltage cable that will run from the electrical grid to a point ten miles out at sea, allowing companies installing wave energy systems easy access to the grid.

OIW's Chandra Brown is frustrated by the lag. "Why did Vestas [a Danish company that is the world's biggest maker of windmills] and all the other European companies get out ahead? Because they were subsidized early and got a corner on the market. Here research hasn't been funded. For the next generation of renewables, we'd like the U.S. to at least be in the game. There will be markets for this stuff when there's a cap on carbon and emissions trading."

————

THE INNOVATIONS THAT ARE IN THE WORKS fall into three broad categories. Each technology carries its own mix of advantages and potential environmental impacts. And most illustrate, once again, the importance of scale in evaluating renewable-energy technologies. A pilot project—whether Finavera's experiment at Makah Bay or Wave Dragon's demonstration project off the Welsh coast, described below—may seem to have a small impact when compared to existing technologies. But a single, 1-megawatt coal plant would not have a large environmental impact either. The concerns emerge at commercial scale, as the combined impact of many generator units potentially affects important ecological processes. Ultimately, alternative technologies must be judged on their impact at the scale of the conventional energy-generating operations they hope to replace.

Shoreline and near-shore devices are relatively easy to install and maintain and do not require underwater cable, but the waves they tap are less powerful because their energy has diminished as they come close to shore, absorbed by the seabed and dissipated through turbulence and friction. A leader in this category is Wavegen, which has operated a grid-connected wave energy plant near its home base in Scotland since 2000. Wavegen was purchased in 2005 by the German water energy joint venture Voith Siemens Hydro Power Generation. Called Limpet (an acronym for "land-installed marine-powered energy transformer"), Wavegen's power plant is centered on a capture chamber—the first was excavated directly out of a rocky shoreline; others have been built into man-made breakwaters. As waves enter the chamber, they compress a pocket of air, which is forced through a turbine at the back; as they leave, they suck the air back out through the turbine, which is specially designed to capture the flow in both directions to create electricity. Using air instead of the moving water itself to spin the turbine blades keeps corrosive saltwater away from the moving parts. This method of capturing energy with rising and falling water is called an "oscillating water column." There is no need to lay a seabed cable—eliminating worries about the impact on bottom habitats and organisms. On the other hand, these facilities have a footprint on coastal lands, where many different kinds of environmental pressures often collide.

Wave Dragon, a Danish-born company that moved to Wales in 2007, uses a completely different technology. The Wave Dragon is a seaborne hydroelectric dam, a large floating barge that stretches out collector arms toward oncoming waves to guide 330 yards of wave front up long, curved ramps; after climbing the ramps, the water flows over the top of the barge and into a reservoir, then drops down through turbines as it returns to the sea. With an expected capacity of 4 to 7 megawatts, the Wave Dragon is the largest onshore wave energy plant in development.

Offshore plants exploit the more powerful waves in deep water. The one AquaBuOY has to beat is the Pelamis from Scotland's Ocean Power Delivery, an articulated "sea snake" that floats partially submerged in the ocean, moored with weights that swing it to face oncoming waves. As its sections are moved to and fro by the waves, hydraulic rams in the joints pump oil to drive generators; a five-hundred-foot snake makes 750 kilowatts of power. Ocean Power Delivery has operated a pilot in the North Sea since 2004 and built its first commercial-scale project (2.25 megawatts) in Portugal; it will add a 3-megawatt project in Scotland in 2008. Since 2002 the company has secured more than $50 million from European venture capitalists, the U.K. government–funded Carbon Trust, and, most recently, GE. Even so, its path, like that of all these new energy ventures, has not been entirely smooth. Soon after installation, the generators in Portugal had to go back for retrofit-ting and repair. According to Finavera and less-biased observers, its immense scale and complexity may prove its Achilles heel.

A THIRD CATEGORY of renewable ocean energy—tidal energy—got a big boost in 2007. While several companies had been working on tidal projects on a small scale, Voith Siemens Hydro announced a quantum leap: the planned construction of a 600-megawatt plant in the South Korean province of Wando. A series of bridge-like structures will suspend turbines, much like underwater windmills, into the current; twice a day, they will rotate to capture the flow direction of both high and low tides. According to Voith Siemens, the jump in scale is made possible by advances in the computer modeling of tides, which enable engineers to do site-specific assessments of different turbine designs and differing arrays of those turbines without having to build or install real equipment. The turbines will be capable of producing 1 megawatt each, an enormous amount compared with other wave energy systems, despite their intermediate size (blades fifty feet in diameter). The reason for the high energy output is location: Wando, whose many small islands create constrictions where ocean water must flow rapidly, has among the fastest currents in the world. Jochem Weilepp, head of ocean energies at Voith Siemens, cautions that the technology is new and challenging; while the company is striving for minimum environmental impact, including an oil-free design, he says the company will not be able to eliminate all problems simultaneously. Plans call for three prototypes to be lowered into the water in 2009, with gradual scale-up to the 600-megawatt installation by 2018. Other companies are considering similarly large-scale tidal generators suspended in the powerful Gulf Stream off the southeast coast of the United States.

The recent advances in tidal energy have triggered a kind of land rush, with companies vying to lock up FERC permits for sites ahead of the competition. Chevron planned to apply to FERC to use tidal power for its offshore platforms in Alaska's Cook Inlet, but Alaska Tidal Energy got to FERC first. Eventually, FERC made them share: each company received a preliminary three-year study permit, and each had to reduce its study area so that the tests would not overlap. In New York City, a fiercer fight has unfolded over the East River between New York–based Verdant Power and Washington, D.C.–based Oceana Energy, which is, as it happens, Alaska Tidal's parent company.

Verdant's cofounder, William "Trey" Taylor, has worked for more than a decade to commercialize tidal turbines. He and his partners first prototyped a design by Philippe Vauthier, an erstwhile jeweler once commissioned by Tiffany & Co. to make a chalice for Pope John Paul II; though they sank $500,000 of personal savings into the effort, a falling out with Vauthier forced Taylor and his partners to move on. They found another design by Dean Corren, a former New York University scientist who in 1986 had figured out how to pitch and twist the watermill blades to overcome one of the knottiest problems in tidal energy: keeping the turbines from stalling as the current slows. Corren had let his patent lapse, but Taylor pulled it and its inventor back into action. In 2002 he built a prototype and a $100,000 catamaran to drag the prototype through the Chesapeake Bay, simulating tides. On the first test, someone forgot to put in a 50-cent cotter pin and the rotor sank to the bottom.

Undeterred, Taylor's team moved its computers and controls into motor homes under the Roosevelt Island Bridge and put their turbines into the East River—a tidal strait—where they continue to operate. Having burned through their own money, they got backing from Matt Klein, who had made an Internet fortune and retired at age thirty-one; he in turn introduced them to Paul Tudor Jones, a billionaire hedge fund manager and chairman of the

National Fish and Wildlife Foundation, who invested $15 million. By 2007 they had six turbines in the river, were powering a grocery store on Roosevelt Island, and had spent $2 million on fish sensors to prove their turbines "don't turn local striped bass into sashimi." Using twenty-four sophisticated "hydroacoustic beam transducers" and an ultrasound system, the sensors monitor the movement of fish—and sometimes birds—as they move through the water. The evidence so far suggests that fish swim around the turbines; next, they plan to test what happens in a larger field of thirty turbines, where fish may not be able to avoid the turbines entirely, but will have to swim between them.

Verdant's rival, Oceana, is at first glance a far more elegant affair. Operating out of a Beaux-Arts building in Washington, D.C., that was once home to banker Andrew Mellon, it is chaired by William Nitze (son of NATO architect Paul Nitze), a Harvard-educated lawyer who spent fourteen years at Mobil Oil before going to work for the Reagan State Department and the Clinton EPA. In 2005 Nitze invested $250,000 in Oceana in return for 20 percent equity. John Topping, president of the Climate Institute, joined him as cofounder.

The Oceana engineering team is a bit more rough and tumble. Ned Hansen, the company's chief engineer, has worked on bunker-busting nuclear weapons at Sandia National Laboratories and on Screaming Squirrel roller coasters. Herbert Williams, who designed the beta version of the turbine, was an Alaska crab-boat captain and served four and a half years in federal prison for conspiracy to distribute cocaine after he designed a fast boat for a Colombian smuggler.

The rival companies have taken very different approaches. Verdant focused first on refining its technology. To answer environmental concerns and secure the permits needed to put its turbines in the water for testing, the company designed a rig that could quickly lift those test turbines out of the water if any problems cropped up. Sure enough, within a few days of putting their turbines into the East River, the blades began to break. Verdant pulled them out for a redesign, using stronger materials and constructing the turbines around a more rigid internal frame. Engineers also redesigned the base that holds the turbines to the riverbed, to afford more resilience against horizontal pressures and greater flexibility in moving turbines from place to place to optimize power output and minimize their impact on fish. After testing at the National Renewable Energy Laboratory, the newest generation of blades was scheduled to be back in the East River in February 2008. Verdant's turbines are much less powerful than those planned for Korea; because of the slower currents and shallower waters in New York, each is capable of producing only 35 kilowatts, about one-thirtieth of Voith Siemens' projected turbine output. But Taylor says the smaller size will make testing and repairs much easier.

Oceana's first efforts were aimed at securing potential test sites. Without at first revealing much about the technology it wanted to test, the company applied to FERC for permission to study a number of prime tidal energy sites, including the currents under the Golden Gate Bridge and a fast-flowing section of the East River, directly upstream from Verdant's Roosevelt Island pilot site. Since FERC's three-year study permits were granted on a first-come, first-served basis, Oceana's were won without a great deal of scrutiny, under the names of seven local-sounding subsidiaries, including Maine Tidal Energy Company and Golden Gate Energy Company.

In September 2006, Senator Orrin Hatch of Utah, where Oceana planned to manufacture its turbines, had written to FERC, urging the agency to grant the company's permits. Oceana asked Utah's other senator, Robert Bennett, to get $1 million into the 2008 federal budget for testing Oceana turbines at the Carderock Division research center in West Bethesda, Maryland, where the U.S. Navy tests model ships in a pool more than a half-mile long. The federal funding did not come through, but the testing was scheduled, nonetheless, for late 2007.

Oceana's innovation is an "open center turbine," based—not surprisingly, given the career of chief engineer Ned Hansen—on linear motor technology borrowed from the roller coaster industry. The open center of the turbine means that aquatic life can swim right through the turbine, and need contend only with the slowly rotating blades near the outer rim. The Oceana turbines feature only one moving part—the rotor itself—and thus are highly resistant to corrosion or mechanical failure. The prototype is just six feet in diameter, although much larger versions are planned for deployment. Compared with other designs, Oceana claims, its turbine is more scalable; the size and shape of its blades are easily varied.

First, however, the engineers want to test and deploy small turbines in tidal inlets in San Francisco Bay, at sites off Maine, Oregon, and Alaska, and in New York's East River. When Verdant discovered that Oceana had applied for a test site adjacent to its own East River turbines, the team was furious. Calling Oceana "claim jumpers," Taylor had Verdant's lawyer file a protest with FERC; it likened Oceana to "dot-com exploiters who seized domain names and held them for ransom." (FERC approved Oceana's test site, despite the protest.) Oceana's general counsel told Bloomberg News that he understood why it might look like Oceana was trying to corner the market, but that in fact it needed to study lots of sites.

Oceana's strategy paid off in San Francisco. In the summer of 2007 the city signed an agreement to collaborate with the company and with PG&E on developing the energy potential of the tides under the Golden Gate Bridge. The utility also announced that it would invest $1 million in Oceana's local subsidiary.

OF ALL THE FRONTIERS BEING EXPLORED and staked by the new-energy prospectors, the oceans and tidal straits may be the most untamed. From Makah Bay to the East River, the regulatory framework is lagging behind the boom. In April 2007, Doug Rader, principal scientist for oceans and estuaries at Environmental Defense Fund, testified before the House Committee on Natural Resources on the need to fix the "fractured system of ocean governance." Two months later, Finavera CEO Jason Bak echoed that argument to the Senate Committee on Energy and Natural Resources. The Minerals Management Service in the Department of the Interior had challenged FERC's hydropower licensing authority on the outer continental shelf, a prime location for ocean power development because the waves are powerful and there is less competition with commercial fishing. Bak urged the committee to confirm FERC's authority, both because his industry had already spent millions complying with its licensing process and because Makah Bay was demonstrating that the FERC process worked even in fiercely protected natural areas. Jurisdictional uncertainty, he told the senators, is "creating substantial regulatory risk for the ocean wave energy industry."

That risk made financing close to impossible to secure. Money rarely flows where the regulatory hurdles are too great. "Getting the technology right is the smallest part of it," says Alla Weinstein, who moved from day-to-day operations for Finavera on to its board of directors, then in early 2008 left the company. "It's the context we're operating in that really matters."

The best way to get the context right, Rader told the House Committee on Natural Resources, is to create a lead regulatory entity that will work with scientists, local communities, and companies to produce comprehensive and predictable rules. Ocean energy can contribute a great deal toward the protection of our atmosphere—without damaging marine ecosystems that are equally vital to the planet's future.

CHAPTER 7

Power from the Earth

A kind of frontier spirit animates this new energy world: a thirst to venture into the unknown; faith that untapped natural bounty, this time truly limitless, waits just over the horizon; even a moral calling, a new kind of manifest destiny that nearly inverts the nineteenth-century aspiration to dominate nature. This time around, technology will not be the harsh master of nature, but its ally.

Nowhere is that frontier spirit closer to the surface than on the actual last American frontier, deep in the interior of Alaska in a little community of sixty-one people called Chena Hot Springs. Discovered in 1905 by a rheumatic prospector, Chena is an eccentric, ramshackle place, hammered together from pipes, wood, and wire salvaged from Valdez and Prudhoe Bay; a resident moose frequently loiters amid the massage huts and piney cabins. It is literally the end of the road, about sixty miles northeast of Fairbanks and thirty-three miles off the grid. Beyond here, bush planes (or sled dogs, but we'll come to that) are the only alternatives. Situated on the "auroral oval"—the latitude (65°N) at which the earth's upper atmosphere produces the most spectacular northern lights—it is a winter destination for Japanese tourists who soak in the waters as the aurora dances overhead, and a summer haven for brides who wed under the midnight sun in a Gothic palace carved entirely of ice. It is also at the cutting edge of geothermal energy, winner of the 2007 "top 100 R&D" projects in the nation (named by *R&D* magazine and the U.S. Department of Energy), and run by a man as outlandish and inspiring as the resort itself.

That ice palace, improbably, is what set proprietor Bernie Karl on his path as a pioneer and proselytizer for clean energy. With his big belly and beard, billed cap and suspenders, Bernie (he is "Bernie" to everyone) is an impresario from another age, bursting with can-do spirit, possessed of a visionary ambition that borders on the nutty, and astoundingly resourceful, patching together world-class projects from recycled junk and ingenuity. One of his heroes is Walt Disney, who inspired Bernie's decision in 2003 to hire thirteen-time world champion ice carver Steve Brice and his wife, four-time champ Heather Brown, to carve the six-room Aurora Ice Hotel out of fifteen thousand tons of ice and snow. Bernie's team embedded refrigeration tubing in an insulated exoskeleton, and circulated glycol cooled by Caterpillar-built diesel generators at a cost of $700 a day.

It didn't work. Within a few months of its January 2004 opening, as the hotel "was melting along with [Bernie's] $20,000 investment," *Forbes* magazine dubbed it "the dumbest business idea of the year. . . . [H]e somehow miscalculated the effect of 24 hour summer sun and 90 degree heat."★ Bernie gamely joked with the reporter, "I had a frozen asset, and I turned it into a liquid asset." And then— he built it again. "Because," he explains, "we'd learned how not to do it." (Constructive failure, it turns out, is a crucial way station for many clean-tech inventors on the way to success.)

Bernie's wife, Connie, whom he met in 1976 when both were working on the trans-Alaska pipeline (she was a bus driver, he a

---

★ *Forbes,* July 5, 2004.

mechanic), told him he had one more shot. So he hired an architect and structural engineer, who eventually shaved the building's weight to one thousand tons, eliminated the use of snow (which is structurally fluid), and redesigned the cooling system to replace the tubing with a big, cold air space between two walls.

Several years earlier, Bernie had hired Gwen Holdmann to figure out how to make power from the hot springs that were Chena's claim to fame. A young space physicist and mechanical engineer from Wisconsin, Gwen had started her own renewable-energy consulting firm. It was the first time, Bernie recalls, that any potential employee had presented him with a three-page description of the job she planned to do; he was so taken with her that he asked her to be his vice president for new development. Gwen was not so sure. "Bernie's a big talker, and I didn't know whether he was for real." But she agreed to begin mapping and analyzing the geothermal resource, secured a $1.4 million grant from the U.S. Department of Energy, and assembled a team of advisers, including Southern Methodist University geophysicist David Blackwell and Dick Benoit of Sustainable Solutions in Reno, Nevada. Together they began drilling holes—some of them thousands of feet deep—to figure out the underlying geology, how deep the water was circulating, and the size and maximum temperature of the reservoir. "Halliburton would have come in and drilled us a deep well for $10,000," says Gwen. "But Bernie didn't have any money, then or now. We couldn't even afford the instruments to get accurate temperature and pressure readings, so I spent $700 and built my own."

Then the hotel melted, and Bernie sprang the news that he had given $50,000 to a nuclear physicist named Don Erickson, who claimed to have invented a three-pressure geothermal absorption chiller (most systems operate at only one pressure), and who had promised to build the first one ever at Chena in two months' time. With the second crystal palace gleaming under the intensifying spring sun, Bernie's calls to Erickson grew frantic. "This thing is going to melt. I'm sending you one-way plane tickets. You have to fly it in tomorrow."

By all rights, that second system should also have failed; numerous experts assured Bernie he was just throwing more money away. Though absorption chillers have been around since the 1800s, conventional physics had set the minimum temperature differential (or delta T) required to make them work at 160°F. With the hot springs water at 165°F and a neighboring creek running at 40°F, Chena had a delta T of just 125°F. Bernie was not phased by the experts' gloom: "A lot of educated people," he says, "learn just enough to know what won't work." (He is not, as a rule, overawed by experts. When Connie needed a $10,000 gallbladder operation, Bernie, who has no medical insurance, brought a medical textbook into the surgeon's office, opened it to a picture of a diseased gallbladder, and said, "You'd better bring me something that looks like this or I'm not paying you." Only when the surgeon handed over Connie's gallbladder did Bernie peel off the hundred-dollar bills.)

So despite the doomsayers, Bernie turned Erickson and Holdmann loose, and sure enough they figured out how to push the existing technologies beyond the edge. (That trick—taking existing stuff and making it go beyond what experts say is possible—is another recurring theme.)

The absorption chiller works, like conventional air-conditioners, by tapping the cooling effect of evaporation; Gwen likens it to the chill you feel at the beach when ocean water evaporates from your skin, drawing out body heat. Instead of using electricity to compress the refrigerant, however, Chena's absorption chiller is the first three-pressure chiller in the world to use geothermal energy.

This is how it works: Hot water from the springs heats a solution of ammonia and water. The ammonia, which boils at just 4°C (40°F), turns into a pressurized vapor, separates from the water, and goes into a condenser, where cold water from Monument Creek turns it back to liquid, still pressurized from the energy it got from the ground. It is when that high-pressure liquid is allowed to expand that it sucks up heat, in this case from a brine that cools the air circulating between the inner and outer walls of the ice "museum," as the structure is now called. The ammonia, meanwhile, is reabsorbed by the water and begins the cycle again.

Though only the size of a big home furnace, the chiller makes enough cold to produce fifteen tons of refrigeration a day (one ton of refrigeration can freeze two thousand pounds of water at 0°C (32°F) in twenty-four hours). That is enough to preserve the vast museum through the longest, hottest summer days: to keep its life-sized ice knights forever jousting on their crystalline steeds, the ice martini glasses brimming with real martinis, the ice chandeliers dangling dozens of individually faceted crystals, and the ice beds solid under their soft caribou hides—all lit with fiber optics to mimic the colors of the aurora borealis. The whole thing is cheesy and yet undeniably beautiful, with bits of subarctic algae, star-shaped crystals, and air bubbles preserved within the luminescent ice. Standing outside as tourists line up to get in, Bernie grins. "The stupidest business idea? *Forbes* can kiss my ass."

That success energized Bernie and Gwen for their next grand scheme: making all the resort's electricity out of the lowest-temperature geothermal resource ever used for power production. (In Iceland, one of the world's major geothermal energy centers, the water is 200°C (400°F). "The stuff they discharge," says Gwen, "is hotter than our water.") They had begun talks with a company about building a customized and expensive plant when they got a call from United Technologies Corporation (UTC) of Connecticut, a $48 billion Fortune 500 company whose holdings include Otis elevator, Sikorsky aircraft, Pratt & Whitney jet engines, and Carrier refrigeration. UTC had heard about Bernie, the caller

said, from the Department of Energy, which had been supporting Gwen's exploration work. Would he be interested in working
with UTC to modify one of its new PureCycle 200 power plants,
developed to make electricity from waste heat at temperatures
between 260°C and 540°C (500°F and 1,000°F), to run on geothermal energy?

UTC had recently spun off its fuel cells and generators into a
new division, UTC Power, and the PureCycle 200 had come out
of brainstorming sessions aimed at reconfiguring into new products
existing technologies in the company's portfolio. Carrier has 69
percent of the world's market share in air-conditioners: how about
running those air-conditioning units backward? In reverse, the
centrifugal compressor would become a turbine: instead of using
electricity to turn the compressor, it could turn the compressor
with steam to make electricity. The UTC engineers found an old
jet engine lying around the company's research center, hooked up a
Carrier unit, pumped in some steam, and in the first test produced
60 kilowatts. A bit more research and development, and they were
testing it on methane flares at landfills. But they needed a more
consistent heat source, which led them to geothermal—one of the
rare renewable sources that is available virtually 100 percent of the
time—and to Bernie. What really swung Bernie UTC's way was
the company's long-term goal: to mass-produce low-temperature
geothermal power plants, cutting the cost from $3,000 to $1,300
per kilowatt installed. Even more than the tightwad Bernie, who
wanted to save the $365,000 he was spending on diesel fuel each
year, the missionary Bernie wanted to spread good works throughout the land.

When the ribbon was cut on the first UTC Chena Power Plant
in the summer of 2006, it was a gala affair. Alaska's political lumi-
naries followed the Fort Wainwright marching band past the goat
pen and organic vegetable garden up to the plant to celebrate.

"We've had a real problem in the past getting partnerships going" for alternative-energy projects, said Senator Ted Stevens. "We've not had such a leader as Bernie." Governor Frank Murkowski also had praise, reported *Petroleum News*, "for the self-described imposter with no formal education who had upstaged the assembled politicians and scientists. 'Most people didn't believe you could take hot water and make a generator out of it,' said the Governor. 'You've proved them wrong.'"[*] Jean Copin of the United Technologies Research Center compared Bernie's adventurous spirit to that of their founder, Elisha Otis, who at the 1853 Crystal Palace Exposition in New York went up ten stories in an elevator and told his assistants to chop the cable to demonstrate the effectiveness of his safety brake. (Bernie himself tells that elevator story often, to explain his deep regard for UTC. "That's the kind of company they are—the real McCoy.")

The Chena Power Plant is even simpler than the chiller. Water from the hot springs is pumped through a heat exchanger, vaporizing a "working fluid" with a very low boiling point; that gas moves through a nozzle at a pressure of 220 pounds per square inch and at twice the speed of sound, spinning a turbine at 15,000 revolutions per minute. The turbine, in turn, spins a generator. The working fluid then goes to a condenser, where it is cooled by Alaska creek water siphoned from a well thirty-one feet uphill. Except for one month when the units were down after a welding spark nearby set off a major fire, the pair of 225-kilowatt "organic rankine cycle" (ORC) plants have worked at full capacity, producing 3 million kilowatt-hours of electricity in 2007, displacing 224,000 gallons of diesel, and saving Bernie hundreds of thousands of dollars.

What is revolutionary about the plant is not the technology, which was familiar, but that a giant company "with the smartest

---

[*] *Petroleum News*, August 27, 2006.

guys in the world," in Bernie's estimation, "were willing to deal with this country mouse" to work out the economics on a small scale. The trick was to invent as little new as possible: of the 171 parts in the power plant's turbine/generator assembly, 158 came off Carrier's existing production line in Charlotte, North Carolina; only 13 new parts had to be manufactured. A Carrier mechanic, says Gwen, would not be able to tell the Chena turbine and generator from the compressor and motor he is used to repairing. By using "the Henry Ford approach," said Halley Dickey, UTC Power's sales manager for geothermal business development, "we cut the cost of power production here from 30 cents to 7 cents a kilowatt-hour, which means a million-dollar plant will pay for itself in a year. What Bernie's built is a money machine."

Connected by satellite and shared monitoring systems, the Chena-UTC team continue to improve the product. UTC team members raised the turbine a few feet because they found that gravity helped reduce foaming as the working fluid dropped back down to the heat exchanger; they also scored the water pipes and added a thick steel cuff on the compressor to increase heat diffusion. And they switched to a cheaper working fluid—R134a refrigerant, which, as Bernie says, you "can buy at Sam's Club"—which allowed the use of cheaper components. (Eventually, they may want to find yet another substitute; use of that refrigerant—common in automotive air-conditioning systems—has itself been challenged as a potent source of greenhouse gases.) The Chena team members added a muffler to deal with the supersonic whine and a dual cooling system: by switching to an air-cooled compressor in the winter, when the ambient temperature hovers around −43°C (−45°F), they reduced the energy needed to run the system in order to get more net power. They also added 3 megawatts' worth of batteries, which Bernie bought from a San Diego dot-com that went bust, for start-up and load balancing.

The enthusiasm at both ends was so great that instead of the usual ten years it takes to bring a new product to market, the Chena-UTC team finished in two. When the fire burned out all the wiring and controls, UTC engineers took vacation time to come to Alaska to rebuild the control boxes, and project manager Bruce Biederman spent his Saturdays on the phone with Gwen, helping her get them going again. And long after Robert Hobbs, director of operations for UTC's Research Center, considered the technology worked out enough to pass it on to product development, the engineers in his division were so fond of the new machine that they wanted to keep working on it. UTC is now working to scale up to a 1-megawatt plant. Bernie has offered to install one at Chena and finish the research and development there.

There are major differences of style. "A big corporation moves slowly, does everything by committee," says Gwen. "We just do stuff. If we screw it up, we fix it." "They're old, stoic, and conservative," says Bernie, "which drives me crazy but is also their strength." UTC's size and history have proved to be enormous assets: "With our engineering and manufacturing expertise, our deep pockets, the fact that we didn't have to start from scratch but could expand an existing production line and our credibility in the market," says UTC vice president Judith Bayer, "we can bring new technology quickly to market." UTC is now building 225-kilowatt power plants for commercial sale, while continuing work on a 1-megawatt unit. The company plans to produce hundreds of the new units each year, which together will generate 100 to 150 megawatts of electricity, enough to supply 100,000 homes—or a city about the size of Toledo, Ohio. More than half of the first year's production has been sold to Raser Technologies, a Provo-based company that holds geothermal leases on 14,000 acres in southwest Utah and will use 135 plants to produce 30 megawatts of power, enough for 30,000 homes. That's baseload power: always available, and completely carbon-free.

UTC's motivation to address carbon emissions began with recognition of its own outsized impact on the world. The company estimates that its products—airplane engines, elevators, air-conditioners, helicopters—generate 2 percent of all carbon emissions worldwide. As an aerospace company, it also knows that any gain in efficiency is worth many times its weight in gold. Since 1997 the company has doubled in size while cutting energy use in absolute terms by nearly 19 percent. Between 1997 and 2006 that cut in energy use meant a reduction in carbon emissions of about 400,000 metric tons, the equivalent of taking 80,000 cars off the road.

UTC hopes to be rewarded for its early greenhouse gas reduction initiatives when a mandatory cap-and-trade program becomes law. Although its power division has just 487 employees (compared to 61,000 at Otis and 41,000 at Carrier), it receives 20 percent of the UTC Research Center budget, an indication of where the company thinks its future profits lie. For its 2006 strategic planning meeting, the company even invited Bernie Karl.

THE VENTURE CAPITALISTS scrambling to get in on solar and biofuels start-ups have largely ignored geothermal power, being more comfortable, perhaps, with silicon chips and biotech than with drill rigs and old-fashioned industrial technology. It is big companies, like UTC and the Israeli giant Ormat Technologies, that are poised to reap the potential, which by 2025, according to the National Renewable Energy Laboratory, could meet between 4 and 20 percent of current U.S. electricity needs. In the fall of 2007, Iceland's Glitnir Bank announced that it would invest $1 billion in U.S. geothermal projects over the next five years and predicted that annual sales of electricity from geothermal sources will grow from $1.8 billion to $11 billion by 2025. The resource requires careful management: to avoid cooling Chena's production wells, Gwen had to develop ongoing monitoring systems to track the water reinjected into the ground.

Heat energy is continually regenerated in the earth's crust, through both the subduction of tectonic plates and radiogenic decay. It is particularly close to the surface along the "ring of fire" where tectonic plates meet, including the entire Pacific coast, where it produces hot rock all the way to the Rockies. A granite pluton— a body of rock formed by the cooling of molten lava—like the one underlying Chena promises a particularly good resource. Granite fractures easily, creating conduits for water that has circulated deep into the crust, and it is often rich in uranium and thorium, which generate heat as they decay. Where underground water comes in contact with deep hot rocks, it can reach 370°C (700°F). At "flash plants," like California's Geysers, water shoots up the wells to the surface, where, released from high pressure, it "flashes" into steam and spins the turbine generator. Following a period of decline in pressure, water is now reinjected into the wells using reclaimed wastewater piped in from cities. After forty years of use, running day and night, Geysers still generates enough electricity for a million people.

Alaska has more geothermal resources than any other state, although none had been developed before Chena. It also has the nation's highest energy costs, ranging from 30 cents to $1 per kilowatt-hour. Though the winds in Alaska are terrific, melting permafrost can compromise the stability of windmills; solar energy is also complicated, given that power needs are greatest in the winter, when the sun scarcely rises. Gwen finds those challenges invigorating: "Alaska is the most energy-intensive state in the most energy-intensive country. So if we can get Alaska to shift, others won't be able to claim it's impossible. And because energy is so expensive, what might be marginal in another part of the country makes sense here. So we can do the R&D to lower the costs. The rest of the country is catching up to us in energy prices anyway." Alaska also has powerful Republican senators, Ted Stevens and Lisa Murkowski, with a large voice in energy policy; both support mandatory federal limits on the emissions that cause global warming.

Bernie has his sights set farther afield, to America's oil-producing regions, which he says could be "the Saudi Arabia of geothermal." In tens of thousands of oil wells, water at temperatures between 120°C and 150°C (250°F and 300°F) comes up with the oil. Considered a nuisance, it is separated and dumped, at a cost of about $4 per barrel of oil. In Texas, in the year 2002 alone, more than 12 billon barrels of hot water were produced, representing a potential geothermal resource of 10,000 megawatts of energy. "That's the equivalent of ten nuclear power plants," says Bernie, "but they're still planning on building nukes and coal." On more than one occasion, he has stood before a roomful of oil and utility executives and told them, "You ought to be ashamed." His partner at UTC, Bruce Biederman, has begun setting up demonstration projects at those Texas wells, putting little Chena power plants on flatbeds and trucking them right to the sites. "There are 500,000 oil wells drilled in the U.S.," says UTC's Judith Bayer, "which means the front-end investment and risk of assaying and drilling [the most expensive steps involved in developing geothermal power] is already done; the permits are already secured. The hot water, in other words, is free fuel."

In October 2007 Bernie got a chance to demonstrate the potential closer to home. The Department of Energy awarded $724,000 to Chena Power to begin using the wastewater produced by Prudhoe Bay oil wells to replace a portion of the 162 megawatts of electricity now generated by the oil field's natural gas–fired turbines. The grant will be matched by funds from Chena Power, UTC, and BP.

Southern Methodist University professor David Blackwell, who continues to work with Gwen on assessing the Chena resource, advocates going further still. A 2007 MIT report he coauthored, "The Future of Geothermal Energy," proposes using advanced oil-drilling techniques to drill down five thousand feet or more, fracture hot dry rocks in the earth's crust, and inject water to "mine heat" that exists along the underground thermal incline everywhere under the earth's crust—if the drills go down far enough. Such heat mines could be drilled near urban centers, Blackwell and his colleagues argue, providing several times the energy those cities need.

FOR BERNIE, WASTE IS A KIND OF SIN and recycling almost a religion; before he was a resort owner, he owned and operated K&K Recycling for twenty-four years. The Chena resort and power plant have beams recycled from an Exxon rig in Russia and fifty-year-old pumps from a Nike missile station. Bernie is looking into buying a 6,500-foot drill rig once owned by Bill Gates on which he hopes to save a couple of million dollars. "The only natural wealth we have is what we can grow or we can mine," he says. "I mined gold for nine years, but then began mining what other people had already mined."

That ethic began early. Bernie grew up in Peoria, Illinois, one of sixteen kids, including five brothers afflicted with muscular dystrophy. He started his first business at the age of eight, in 1960, with a lawnmower he bought on credit for $108 from Sweetman's Hardware Store. He paid for it by mowing the Sweetman lawn at the top of a big hill twenty-two times. Then he bought ten more. Two years later he had ten kids working for him mowing lawns, was on his bike delivering the *Chicago Tribune* early every morning and the *Peoria Journal Star* each night, and pulled a wagon around town daily collecting the previous day's papers, storing them in his parents' garage, then loading them into his uncle's coal truck to sell for 25 cents per hundred pounds. When he sold a Stark Brothers fruit tree to his third-grade teacher, Mrs. Webbler, it was the start of yet another business. Equipping his wagon with batteries, an inverter, and an electric hedge clipper, he offered full-service landscaping to his neighbors. From his Dad, who worked for Caterpillar for forty-three years but spent every free moment expanding their house, Bernie learned plumbing, electrical work, bricklaying, and carpentry. But it was from his brothers in wheelchairs, he says, that he got his drive. "It is because of them that I do not dwell on bad things; I work on good things. One of my brothers can't swat a mosquito. But he took a Kubota mower, put a trailer on it like the Popemobile, and runs it with his mouth, mowing acres of grass."

Bernie saw those same intrepid qualities in Gwen. When she showed up at Chena, she had only recently moved out of a squatter's cabin on an abandoned gold claim in the woods, where she had lived alone for a year. She had come to Alaska the day she graduated from college; "like a compass," she says, "I'd always had a pull to the north." Her fifth-grade science fair project had been on the solar wind, which creates the aurora borealis; as a graduate student in space physics at the University of Alaska, she returned to that obsession, mapping the aurora with a 30,000-volt laser that could track the movement of atoms in the upper atmosphere. She also began collecting and training sled dogs. At age twenty-one, with twenty dogs and no money, she had stumbled upon the 120-square-foot abandoned cabin, a half mile from the nearest road, which had a sod roof and root cellar but no water or electricity. To get water in the winter, she had to snowshoe to the creek and chop a hole in the ice; to stay warm, she had to down trees and split untold cords of firewood. Food and propane for gas lamps had to be hauled in on a sled, with Gwen usually doing the pulling. "Most of us spend our lives not thinking about where you get all that stuff. I felt like I learned the true value of water, the true value of warmth and light. How much work does this heat require? I knew exactly: if I want to be warm tonight I have to chop wood for an hour."

Frustrated by the abstract nature of her graduate research and by university politics, Gwen left school and began her consulting firm, Your Own Power. After a year of squatting, she purchased land from the state and built her own cabin, where she lives with her husband, Ken, who finished seventh in the 2007 Iditarod. (Though she worried the building department would restrict her dogs, her construction permit prohibited just four things: a nuclear power plant, a hazardous waste dump, a prison, and a strip club.) Most of the power for their house and Jeep comes from solar panels and used vegetable oil.

In March 2001 Gwen finished the Iditarod, mushing her dogs 1,161 miles from Anchorage to Nome on the coast of the Bering Sea, spending twelve days and nights crossing mountains, frozen rivers, and tundra, maneuvering treacherous precipices in blizzards and howling winds. In 2003 she won the third-biggest dog race in the world, the 430-mile Wyoming Stage Stop. In February 1998 and again in 2004 she finished the most grueling race of all, the Yukon Quest, 1,000 miles along the historic Gold Rush route from Whitehorse in the Yukon Territory to Fairbanks, riding and running alongside her dogs and muscling the sled out of trouble through unremitting darkness, temperatures of forty below zero, gale-force winds, open water, bad ice, and—most terrifying—a head-on collision with a moose. Moose are terrified of dogs (their only predator, apart from humans, is the wolf) and have been known to charge and stomp them to death when alarmed. Nearly as terrifying was a night during the Quest when she stopped to "snack the dogs" and lie on her back in the snow to watch the aurora, which was particularly beautiful that night. Her dogs, suddenly restive and whining, roused her from her reverie to alert her to a pack of wolves in the woods just eight yards away. For the next fifteen miles, she could see the wolves darting in and out of the trees, shadowing her team. Now Bernie, who chides himself for being so fat that he puts out "twice as much carbon dioxide as most people," has decided he is going to run the Iditarod—"with one hundred chihuahuas."

Bernie's other schemes are nearly as farfetched. "People tell me I'm thinking outside of the box. But hell, I've never been in the box. And I'm never going to get in it."

By the end of 2008 he intends to use no hydrocarbons at all at Chena Hot Springs Resort. The next step on the way to that goal is construction of a 400-kilowatt biomass power plant, on which he's again partnering with UTC. Although his mother still raises corn, Bernie is deeply scornful of corn ethanol. "You can't burn food— it's just stupid. And with all the subsidies, I figure its real cost is about $11.50 a gallon. How unconscionable that Congress would dump so much money into it. I asked Senator Stevens why they did, and he said, 'It's 28 votes, the strongest lobby in the nation.'"

What Bernie proposes to burn in his plant are the fast-growing willows that cover much of the Alaskan interior. "Sweden's at the same latitude, and they use willows. You can get 8,000 BTUs per pound, with no irrigation. Two hundred acres would power a whole village. Hell, a moose grows from sixty to two hundred pounds in a year eating nothing but willows. They've got 30 percent protein in the top eighteen inches, and good hay has only 18 percent protein. And tannic acid on the bottom so the moose only eats the top and doesn't kill it." (As Bernie gets increasingly wound up by his own vision, Connie, who is standing nearby, interrupts gently: "I'd verify all that.")

Next on the list is what Gwen calls "Bernie's hydrogen kick." Gwen's analysis of the Chena hot springs suggests that it might ultimately support as much as 5 megawatts of generation. Since the resort uses less than half a megawatt, Bernie is using the leftover electricity to make hydrogen, with which he hopes soon to power all the resort's vehicles. Gwen acknowledges that they could provide a model for the use of Alaska's vast supply of "stranded renewables." The Aleutians, for instance, are a string of volcanic islands in the Pacific's ring of fire that project westward toward Russia. They are a world-class geothermal resource far from major human settlements and power lines, but adjacent to the major international shipping lanes known as the Alaska Marine Highway. Geothermal power plants there could be used to make hydrogen to power merchant vessels.

Plan number three is to commercialize the absorption chiller, which by pushing the lower limit of the delta T has opened up the potential for low-grade waste heat resources around the country. Most Alaska villages still use noisy diesel generators to supply all their electricity needs. Although they use the jacket water that cools their diesel generators for space heating in the winter, they dump that hot water out in the summertime. Instead, says Gwen, they could use the waste heat to run a chiller, make ice for the local fishery, and get a higher-quality, higher-value product.

In Chena's hydroponic greenhouses—already a seven-thousand-square-foot oasis of productivity, with garlands of tomato vines heavy with fruit and heads of lettuce that could win a county fair—Bernie provides a testing ground for numerous agricultural projects run by the University of Alaska. Ultimately, he says, he wants to prove that forty acres can feed six hundred thousand people. He will soon replace the commercial growing medium with ground glass, from the bar's empties. Asked "How do you know it's going to work?" he shrugs. "How do I know it won't?"

Bernie's most important contribution may be as a kind of old-fashioned, plainspoken country preacher. "There's a cost for living on this earth," he says. "We either pay now or our grandkids pay later, and I believe we should pay as we go. We've screwed up our atmosphere because everyone's had a free ride. I believe in users' fees. If you want to trash the atmosphere, you got to pay the price. I also think you should be rewarded for doing things right."

Bernie has "almost single-handedly energized the sleepy geo-thermal industry," says Craig Walker of UTC. He has insisted that none of Chena's innovations are proprietary—that they should be widely shared. And though the stream of visitors to his remote out-post rarely seems to slow, he never seems to tire of it. Dan Driscoll of the Power Enhancement Group of Reno, Nevada, comes by to see the power plant for possible use at gold-mining operations in China and South Africa, and Bernie walks him through. An hour later, two members of the Alaska House of Representatives drop by, and he does it all over again, this time with an *Alaska Magazine* writer in tow, while a bride and her bridesmaids crunch through the gravel nearby. *Popular Mechanics* is on the phone, and an hour later, Discovery TV. At 10:30 at night, when yet another writer shows up, Bernie's out there again, ready to talk till dawn.

THE BUSH ADMINISTRATION budgets for fiscal years 2007 and 2008 requested no research funding for geothermal energy, saying that it was a "mature technology" with no more need for federal money.★ That judgment frustrated proponents of geother-mal power. As UTC's Judith Bayer says, "It's no more mature than wind, gas, or oil, which still get plenty of federal R&D funding." Even the flawed subsidies that have been used by Europe to spur

---

★ Despite the administration's argument, Congress earmarked $5 million for geother-mal projects in fiscal year 2007 and was expected to allocate about $33 million for fiscal year 2008.

renewable energy have had an effect, she notes, getting the European Union out ahead in developing technology while the United States has lagged behind. And with the European cap-and-trade system now in full swing, the gap will only grow.

The administration's timing, Bayer says, was particularly perplexing. UTC had just produced the new Chena power plant with help from Department of Energy funds. MIT had issued its report on heat mining, suggesting that enhanced geothermal technology could fill a significant portion of the country's energy needs. And the U.S. Geological Survey had announced that, thirty years after its last assessment of the nation's geothermal resources, it would issue a new report in 2008. In its 1978 survey the agency had identified potential geothermal resources that could produce 23,000 megawatts and estimated that "undiscovered resources" could generate another 127,000 megawatts. The report was based on technology and assumptions now long out-of-date, but even then, its authors believed they might have underestimated the resources.

Soon after the proposed Department of Energy budget became public, Bernie saw Senator Lisa Murkowski. He recounts the following exchange:

"Who's the idiot I should talk to who decided not to include geothermal in the mix?" he asked.

"Well, that would be the president," Murkowski replied. "But you know what? He would enjoy you, and I'm going to call the White House to arrange a meeting."

George W. Bush has a geothermal heat pump at his Texas ranch, so Bernie expects a sympathetic audience.

Bernie has one last plan cooking at Chena Hot Springs. Not far down the road from him, UCLA professor Alfred Wong runs HIPAS, the High Power Auroral Simulation facility. The facility contains several powerful instruments, including an extremely high-powered laser from Lawrence Livermore National Laboratory, which HIPAS uses to measure phenomena in the arctic ionosphere (upper atmosphere). Wong has ionized dust particles in the ionosphere, creating a thirty-mile aurora that he can make change colors and dance across the sky. Bernie has decided he wants to make an aurora on demand for his Japanese tourists, using, of course, geothermal power. ("Is he on that again?" Connie asks, strolling by.)

Wong is also experimenting with using the earth's magnetic fields to eject carbon dioxide from the atmosphere so it no longer warms the planet—an idea discussed in Chapter 10. And Wong wants to make weather with his "ion driver," an array of magnets charged by the sun. Has he really made it rain? Bernie smiles: "In his mind he has."

For a dreamer like Bernie, that means he's most of the way there.

CHAPTER 8

Reconsidering Coal

In late February 2007, as Fred Krupp was preparing to board a plane, he got a call from former EPA director William Reilly, now a partner at Texas Pacific Group (TPG), one of the nation's largest private equity firms. Reilly told Fred that TPG, Kohlberg Kravis Roberts (KKR), and Goldman Sachs were planning what was, at the time, the biggest leveraged buyout in history. They would offer $45 billion for TXU, the largest energy provider in Texas, which at that moment was seeking fast-track approval to build eleven new coal-fired power plants. Fred had to interrupt Reilly to send his cell phone through the X-ray machine. When it came out the other side, Reilly finished what he had to say. "We won't do the deal," he told Fred, "unless you and the Natural Resources Defense Council help us come up with an acceptable climate plan."

Business as usual ended for coal in 2007, as clearly as it had ended for apartheid in 1984. In that transformative year, with nearly the entire world condemning South Africa's racist regime, money-center banks ceased all business in that country. The TXU deal marked a similar breaking point: America's biggest private equity funds realized that they had to be on the side of reducing—not increasing—atmospheric levels of carbon dioxide.

Coal is the most plentiful—and dirtiest—of all fossil fuels.

Recoverable reserves still in the ground could continue to supply electricity worldwide for nearly a hundred years. U.S. coal alone contains more energy than the oil stored in Saudi Arabia's vast reservoirs;[*] Russia, China, Australia, and India also have considerable coal reserves within their borders, and all see those reserves as key to their energy future. But coal plants also emit more global warming pollution than any other energy source.[†] Depending on the technology and type of coal burned, a coal plant emits 1,600 to 2,100 pounds of carbon dioxide for every kilowatt-hour of electricity—more than double the emissions of a combined-cycle natural gas plant. If the world continues along its current course, the U.S. Department of Energy predicts, global coal consumption will almost double by 2030. The potential consequences for the environment are sobering to contemplate: if burned in conventional plants, carbon dioxide emissions from coal over just the next twenty-five years will exceed total coal emissions from the last two and half centuries.

Given how abundant coal is, and also how much damage it can do, two distinct trends have emerged in recent years, both of which will change the energy equation considerably.

On the one hand, coal's longtime dominance is being chal-

---

[*] Until recently, the U.S. Department of Energy estimated that the United States had several hundred years of usable coal reserves. Newer studies suggest that as little as 5 percent of the 1,700 billion tons in the ground may be recoverable through mining, which would put recoverable reserves at about 85 billion tons, or seventy-seven years at the current consumption rate. It should be noted, however, that assessment of reserves is highly contingent on assumptions made about the cost and availability of technology and projected prices; world reserves of oil, for example, have increased as technological improvements and higher prices have made it possible to recover previously inaccessible or uneconomic reserves.

[†] Burned conventionally, without modern pollution controls, coal also emits mercury, a powerful neurotoxin; sulfur dioxide, which causes acid rain; oxides of nitrogen, which cause smog and acid rain; and particulates that contribute to respiratory and cardiac illness. Most U.S. coal-burning plants have installed varying degrees of controls for sulfur, nitrogen, and particulates.

lenged in many quarters, largely because of the high probability that coal-fired power will become less cost-effective as the costs of curbing carbon dioxide are factored in. Nearly two-dozen coal projects have been canceled since early 2006, according to the National Energy Technology Laboratory (NETL), a division of the Department of Energy; between June and August 2007, analysts at Morgan Stanley, Citigroup, and Goldman Sachs downgraded the stocks of coal-mining companies, citing the evolving regulatory climate. In October 2007, in a move that may well have nation-wide impact, the Kansas Department of Health and Environment denied a permit for two 700-megawatt coal-fired plants in that state solely because of the global warming gases they would emit. The *Wall Street Journal* noted "how quickly and powerfully environmental concerns and the costs associated with eradicating them have changed matters for the power industry."★

At the same time, it is clear that for the next few decades at least, coal simply has to be part of the energy equation; it provides roughly half of all U.S. electricity, and up to 80 percent of electricity in areas like the industrial Midwest. A number of start-ups, and a few of the biggest companies in the world, have therefore begun to focus significant resources on reinventing coal. Some are working to increase efficiency—extracting the maximum energy possible from each ton of coal. Others are working on scrubbers to retrofit onto existing plants to capture the carbon dioxide from flue gases, and on finding safe and long-lasting ways of storing that carbon deep underground. At the furthest edge, innovators are working to fundamentally remake the way coal is used—with advanced methods for turning it to gas before it is burned—which also requires resolving the challenges and uncertainties attached to "sequestering" the carbon dioxide in the ground.

★ *Wall Street Journal*, July 25, 2007.

JUST A YEAR before Fred's airport conversation with Bill Reilly, TXU had refused to consider alternatives to its planned construction of eleven new coal-burning power plants, even though numerous studies had made it clear that statewide investments in energy efficiency would have rendered the new plants unnecessary. If built, those new plants would have added 78 million tons of carbon dioxide to the atmosphere each year, more than the emissions of twenty-one states. With local partners, Environmental Defense Fund had filed suit in federal court to stop the plants. It also had gone to Wall Street to outline for potential investors the substantial risks in acquiring such an enormous carbon burden, even as Congress was debating several global warming bills—none of which would have guaranteed TXU free allowances for the new emissions (allowances that would permit them to emit some level of carbon without penalty).

Texas regional director Jim Marston, the lawyer who had been leading Environmental Defense Fund's campaign to stop TXU, was headed to a public hearing on the plants when Fred called, asking him to get on a plane to San Francisco that day to meet with TPG and KKR. He swore Jim to secrecy. "I literally didn't even tell my wife," Jim says. "I could only tell her that I loved her and that I'd be back." With Fred and David Hawkins from the Natural Resources Defense Council consulting by phone, Jim negotiated for seventeen hours, from eight o'clock in the morning until one o'clock the next morning. By the end, the buyers had agreed to scrap eight of the eleven new plants in Texas, as well as new coal-fired facilities under development in Virginia, Pennsylvania, and Georgia. They had agreed to cut TXU emissions to 1990 levels by 2020, to double the company's investment in both renewable energy and efficiency, and to support a mandatory federal cap on carbon—a complete turnabout for a company that had been one of the nation's noisiest opponents of legal limits. The buyout executives then flew to Austin to confirm their plans with Texas Governor Rick Perry and other state leaders. "Anyone doing an energy investment in today's situation has got to be sensitive of the change in the attitudes of the ... country and particularly the attitudes of Congress," an unnamed participant in the negotiations told the *Washington Post*. According to the paper, the TXU deal could become "a landmark in the battle over climate-change policy."★

Utilities still have plans to build more than a hundred new coal-fired power plants in the United States over the next few years, however. And until there are legal limits on carbon emissions, many will continue to opt for the cheapest and most polluting plants. Those new plants will last some fifty years, committing us to another half-century of emissions. And the fifteen hundred plants already producing power in the United States—representing trillions of dollars in capital investment—will not vanish anytime soon. It is the collision between that huge and growing dependence on coal and the recognition that the planet cannot bear the additional carbon burden those plants will produce that is spurring innovation in how we use coal—although in the absence of legislation, the innovation remains at a scale utterly unequal to the challenge.

NEARLY ALL COAL PLANTS in the world today burn coal that has been pulverized to a powder the consistency of talcum and then blown into a boiler. The vast majority of these pulverized coal plants are "subcritical"—that is, they operate at relatively low temperatures and pressures and convert about 35 percent of the coal energy into electricity. More advanced plants operate at "supercritical" temperatures, typically between 550°C and 590°C

★ *Washington Post*, February 26, 2007.

(1,025°F and 1,100°F), pushing efficiencies to just under 40 percent; "ultra-supercritical" plants, which run above 590°C, can get more than 40 percent of the energy out of the coal. That means less coal—and therefore reduced emissions—per kilowatt-hour. Replacing a 37 percent efficient subcritical plant with a 47 percent ultra-supercritical plant would reduce carbon dioxide emissions per kilowatt-hour by 20 percent, according to Alstom, the world's second-largest producer of generation equipment. In order to achieve efficiencies close to 50 percent, Alstom and several other companies are trying to push temperatures to 760°C (1,400°F) or beyond by developing new metal alloys and ceramics that will not warp or disintegrate in extreme heat and under high pressures. Other companies are using advanced artificial intelligence techniques to optimize the operation of the aging fleet of pulverized-coal plants. NeuCo, for instance, a Boston-based company with offices in Beijing, develops software based on neural networks and fuzzy logic.* With continuous input from sensors throughout the plants, NeuCo's software detects patterns over time and "learns" to make real-time adjustments (in air flows and pressures, for instance) to maximize energy output.

But stabilizing the climate will require far bigger cuts in coal emissions than these step improvements can achieve. Ultimately, emissions must be reduced to nearly zero. For Alstom, based in France and already operating within Europe's carbon constraints, the most promising route to near-zero emissions from coal power plants lies with a process invented by Eli Gal, the Israeli-born chemical engineer introduced in Chapter 1 who at General Electric two decades ago made quantum improvements in sulfur dioxide scrubbers. A lean, vigorous man who still carries shrapnel in his leg from the 1973 Yom Kippur War, Gal is confident that the technology for

---

* Fuzzy logic allows more than a simple yes or no answer to a question; it incorporates the concept of partial truth.

carbon capture—cleaning carbon dioxide out of flue gases—will follow a similar trajectory. "It will become so cheap and simple that plant operators won't notice it," he says. "It's not rocket science."

Perhaps. But reining in carbon dioxide will be more complicated, not least because of the daunting volumes. A 500-megawatt coal-fired power plant produces about 50,000 tons of sulfur dioxide each year. But that same plant produces *3 million tons* of carbon dioxide. If operators at that plant were to use the same strategy for carbon dioxide as they did for sulfur dioxide (reacting the acidic carbon dioxide with a base like sodium hydroxide to make sodium carbonate), they would have 10 million tons of sodium carbonate to get rid of. That's 10 million tons to dispose of—for one plant, for one year.

The answer, Gal realized, was to find a reagent that would be "regenerative"—that is, after grabbing the carbon dioxide and separating it from the other gases, it would have to release the carbon dioxide for sequestration, recycle back into the process, and start over again.

The idea of using ammonia as that reagent has been around a long time. When it was first explored a century ago, inventors were working the other way around: using acidic carbon dioxide to get rid of ammonia (a base), which floated out of gas lamps and filled Victorian streets with its acrid smell. More recently, the Department of Energy had resumed research on binding ammonia and carbon dioxide. Gal had dismissed such experiments as nonsense. At GE, he had worked on sulfur scrubbers that used ammonia; the carbon dioxide in the gases had ignored the ammonia and passed right through.

Still, Gal kept thinking. He left GE to return to Israel, where he spent seven years working for a company that made chillers and desalinators. That's when it hit him: ammonia (or better yet a slurry of ammonium carbonate and water, which is safer to work with and reduces the risk of free ammonia escaping into the air) might work to capture carbon dioxide if the flue gases were cooled first. He returned to the United States and resumed his research. His son David, then an MBA student at Stanford, found him books in the library that had not been checked out for eighty years.

Gal's technology operates on the same basic chemical principle that keeps soda bubbly when it is cold—and flat when it is not. First, the flue gases in a coal-burning plant are cooled to about 5°C to 10°C (40°F to 50°F), at which temperature the ammonium carbonate reacts with the carbon dioxide to form ammonium bicarbonate. The ammonium bicarbonate then goes to a regenerator, where it is heated back up to 120°C (250°F)—using waste heat from the power plant—releasing the carbon dioxide in a pure stream that can be injected into the earth. Though the emissions still need to be scrubbed first of sulfur dioxide, nitrogen oxides, and mercury (which would otherwise hook up with the ammonium carbonate before the carbon dioxide ever got to it), the chiller provides backup scrubbing for any of the pollutants that slip through. As the gases are chilled below the dewpoint for sulfur oxides and mercury, they precipitate out—just as humidity in the chilly early-morning hours turns to dew—into solids that can be easily removed.

Almost none of the hardware for Gal's system had to be newly invented. "We've been building absorbers for years, and the heat exchangers and chillers are standard industrial products," says Bob Hilton, director of business development for environmental control systems at Alstom, which now holds exclusive license to the technology. "What's new is the process: the temperature regime, how we integrate heat from the plant."

This "chilled ammonia" process is not the first to capture carbon dioxide from coal-plant flue gases. The most developed carbon-capture technology on the market uses amines, the nitrogen-based molecules that are the building blocks of amino acids, to bind the carbon dioxide. The problem is how much energy and money that costs: once bound to the carbon dioxide, amines do not want to let go. Separating the two so that the amine can be reused and the carbon dioxide can be pumped into the ground uses more than 30 percent of the power plant's total energy output. That, in turn, pushes up the cost of electricity by more than 50 percent. Newer amine technologies promise to reduce this parasitic load below 20 percent, meaning the cost of electricity will rise by only 44 percent, including sequestration, but that is still a steep increase.

The ammonium carbonate slurry, because it can be made more concentrated than amine solutions, holds twice as much carbon dioxide per volume and needs less than half the heat to regenerate. (Although the ammonia process does require extra energy for the chilling, Alstom claims its net energy use will be lower.) Regeneration from ammonium carbonate also occurs under high pressure, reducing the additional energy needed to pressurize the gas to pipe it to reservoirs under the ground. When fully developed, says Hilton, the chilled ammonia process will use just 10 to 15 percent of the energy output of the power plant, adding about 25 percent to the cost of electricity. Put differently, it will cost about $20 per ton of carbon dioxide removed, plus another $5 to $10 for transportation and sequestration—less than the price at which carbon futures were trading in Europe at the end of 2007. With a U.S. cap, a domestic coal plant that installed such a system could sell excess reductions for a profit.

Gal had been working on the chilled ammonia process at SRI International, a nonprofit research institute in Menlo Park, California, when he and Alstom began talking. "A bunch of us had known Eli for more than twenty years," says Hilton. "Air pollution control is a small world." Alstom built a quarter of all the power equipment now operating in the world; Hilton says the power industry cannot afford to build enough new plants to achieve meaningful carbon dioxide reductions in the near term, so the company is focusing hard on how to improve existing facilities. "We're very concerned about maintaining that fleet, which is going to remain the workhorse till at least 2030," says Hilton. "There's just no way that we have the money as a society to dismantle and replace it, which is why we're so focused on these retrofit technologies."

In 2007 Alstom and Gal began construction of an $11 million pilot scrubber for the flue gases from 5 megawatts at the We Energies plant in Pleasant Prairie, Wisconsin; the Electric Power Research Institute gathered support for the project from more than thirty U.S. utilities. The pilot is designed to capture about a hundred tons of carbon dioxide a day, about 1 percent of the 600-megawatt boiler's total emissions. Before the pilot could be switched on, however, news leaked of an unreleased report from the National Energy Technology Laboratory questioning whether chilled ammonia actually improved on amine scrubbers: both processes, its preliminary findings suggested, would use about a third of the plant's energy, increase the cost of electricity by 50 percent, and cost about $30 per ton of carbon dioxide removed. A spokesman for the national lab later downplayed the findings, saying the information was quickly made moot by improvements in the chilled ammonia process. But the report stirred up coal's many critics, particularly those suspicious that the promise of this technology was being misused to justify building more pulverized-coal plants. Eric Redman, a lawyer who represents several companies working on technologies to "gasify" coal (see below), calls pulverized coal a "horse and buggy technology." Though he concedes that the existing fleet must be retrofitted to cut emissions enough to make a global difference, he argues that companies with big stakes in conventional technology are hyping the unproven chilled ammonia technology. "It's almost immoral," says Redman. "It's like telling someone you can keep smoking because we might come up with a cure for cancer."

In 2008, if all goes well, Alstom will install a demonstration scrubber at the 1,300-megawatt Mountaineer Plant in West Virginia. Mountaineer is owned by American Electric Power (AEP), one of the nation's largest utilities. AEP makes 75 percent of its electricity from coal-fired plants and in most years is the nation's number one industrial emitter of carbon dioxide.* It will spend about $80 million on the Mountaineer project and plans to capture and inject 100,000 tons of carbon dioxide a year (about 1 percent of total emissions) into natural underground "saline aquifers" that the Department of Energy is now testing for their long-term storage potential. As deep as ten thousand feet, these porous rocks are like a salty wet sponge, able to soak up large quantities of carbon. Among the questions being investigated is whether these aquifers are sealed by a caprock sufficiently impermeable to help keep the carbon underground.

By 2011 AEP plans to build a commercial-scale system, at a cost of some $325 million, on a 450-megawatt unit at its Northeastern Station plant in Oologah, Oklahoma. That scrubber will capture 1.5 million tons of carbon dioxide a year, about half the unit's total emissions, which the company hopes to sell into the existing market for use in enhanced oil recovery: injected into declining oil fields, the carbon dioxide helps push out remaining petroleum. (While carbon dioxide from natural sources has been used for years

---

* Late in 2007, AEP agreed to spend $4.6 billion cleaning up conventional air pollutants such as sulfur dioxide and nitrogen oxides in a settlement with the U.S. government, states, and environmental organizations. The settlement was aided by a unanimous Supreme Court decision earlier in the year, in *Environmental Defense et al. v. Duke Energy*, rejecting one of the power industry's central defenses against legal claims that clean-up is required. As the *Washington Post* reported on October 9, 2007, "The justices' unambiguous ruling made it clear to utilities that . . . they might be forced under law to curtail their emissions from aging coal-fired plants."

in the oil industry to increase production, no effort was made to keep it from escaping over the long term. A carbon dioxide storage project would aim to keep the gas underground for centuries.) Alstom is also running pilots in Sweden and Norway on oil and gas turbines. The company's far-ranging efforts have made an impression on Wall Street: while other companies linked to the coal business were seeing their stocks downgraded, Alstom's stock was upgraded by Morgan Stanley, which specifically cited its strong positioning in clean coal.

Alstom also has other experiments in the works. "We never approach anything as the definitive technology," says Hilton. "You never know what's going to be the best longer term, or how many different options you'll need. For sulfur dioxide, we sell nine different scrubbers." With Toshiba, Alstom is exploring cryogenic processes: the flue gases are chilled to temperatures low enough—40°C below zero (−40°F)—to make the carbon dioxide precipitate out all by itself, into dry ice. Defrosting it results in a pure gas stream that can be injected into the ground for long-term storage.

In addition to these "post-combustion" innovations, some companies are experimenting with the front end of the boiler—developing new ways of burning pulverized coal. For instance, rather than burn coal with air, which is mostly nitrogen and therefore produces lots of nitrogen oxides, which must be separated from the carbon dioxide, "oxycombustion" boilers burn the coal with pure oxygen and therefore get almost pure carbon dioxide. The problem is the energy required to produce the oxygen: the most common process requires chilling air down to −180°C (−300°F) to strip out the nitrogen. Alstom is also working on boilers that can handle large amounts of biomass such as wood chips or straw mixed in with the coal, reducing net carbon emissions.

What has moved the big power companies to explore these

many alternatives, says Gal, is the expectation of regulation. "As we saw with sulfur dioxide, this stuff has to be regulation driven," he explains. "The market has to demand it. The power industry doesn't go beyond what it's required to achieve." Indeed, even as AEP was committing to the chilled ammonia demonstrations, and its supplier (Alstom) was saying that scrubbers plus carbon capture and sequestration would increase the cost of electricity by no more than 25 to 30 percent, AEP CEO Michael Morris was telling the *Wall Street Journal* that emissions curbs would drive up power prices by 50 percent or more. He also acknowledged to the paper that he was less focused on the science than on the "political science" of what Congress intends to do.★

Amy Ericson, who manages strategy for Alstom's energy and environment systems, believes her company is gaining competitive advantage from the European carbon cap. It has "really pushed us technologically. More than half our new technology projects are originating in Europe." Hilton, who says he has gotten his share of grief from friends for spending so much of his career in Europe, has not given up on the United States: "It's true that nobody buys pollution equipment until they're forced to. But once everyone's regulated, they'll remove what the reg says they have to remove, and they'll all be competing on the same level again." In the end, Hilton believes, the United States will "take its leadership role"— act, that is, to set clear, predictable rules on carbon cleanup. "Our clients are preparing for what they consider the inevitable," he says. Morris, of AEP, says he wishes Congress would act sooner rather than later.

The urgency is hard to overstate. Over the decades, Hilton explains, Europe and the United States "have gone back and forth as to who steps up first—on particulates, on sulfur dioxide. And

---

★ *Wall Street Journal*, July 12, 2007.

each time, China has followed. But not until then: they expect the more advanced countries to move first." The same is true this time round. China, which makes 80 percent of its electricity from coal, is adding a new 500-megawatt coal-fired plant *every four days*. It now consumes more tons of coal each year than the United States, Europe, and Japan combined. In 2006 it passed the United States to become the world's largest emitter of carbon dioxide, contributing 20 percent of total emissions worldwide.

AS IN SO MANY of the new energy technologies, biology is playing a role in reinventing coal. One of the most cutting-edge technologies for grabbing carbon out of smokestack gases is modeled on physiological processes evolved over millions of years and used by virtually every organism on earth. The scientist at the fore of this research in the United States is neither a chemist nor an engineer, but a neurobiologist who has taught at Harvard Medical School and the Brain Research Institute in Zurich.

Like Isaac Berzin, the chemical engineer who founded Green-Fuel, Michael Trachtenberg did his early work on carbon dioxide removal for NASA: building carbon dioxide–capture systems for spacesuits and for the ship being developed for the first manned mission to Mars. His search for a recyclable method for grabbing carbon dioxide out of the space capsule led Trachtenberg back to his own clinical research on the role of glial brain cells in epilepsy and head trauma. Glial cells are what he calls the "housekeepers" of the brain. While neurons are firing and transmitting information, the glial cells support them, bringing in nutrients, including oxygen, and carrying away pathogens and carbon dioxide. Since 1967, Trachtenberg had been investigating a mystery: given the large distances in the brain between neurons and blood vessels, it seemed impossible that the glial cells could scavenge carbon dioxide fast enough to prevent a dangerous buildup. Increased carbon dioxide would increase acidity, which in turn would change the "excitability" of the neurons, causing them to fire at the wrong time.

Nature clearly had solved that problem. Trachtenberg guessed that, given its "parsimonious nature," evolution had borrowed the same high-speed carbon dioxide shuttle used in nearly every plant, animal, and bacteria on earth: an enzyme called carbonic anhydrase. Plants use carbonic anhydrase to help them absorb carbon dioxide from the atmosphere. Mammals use it to help them move carbon dioxide from tissues into capillaries and from there to the lungs to be expelled from the body. One driver of that movement is a simple pressure differential: a gas will always flow from a region of higher concentration and pressure to one of lower pressure. But the second driver—the one that homes in on the carbon dioxide—is the carbonic anhydrase contained in each of our cells. The enzyme catalyzes a reaction that turns carbon dioxide to bicarbonate, which can be carried in large volumes in the blood plasma to the lungs. There the reaction is reversed; the bicarbonate turns back to carbon dioxide gas, which is exhaled. For both uptake and release, the enzyme accelerates the reaction a millionfold. (And oddly enough, adds Trachtenberg, human exhalations have about the same concentration of carbon dioxide—4 percent—as a power plant fired by natural gas.)

Confirming that glial brain cells indeed contained large amounts of carbonic anhydrase gave Trachtenberg his idea: to insert that enzyme into an artificial liquid membrane that could keep capturing the carbon dioxide in the spaceship all the way to Mars and back, dumping it out into the empty reaches of space. He was not the first to recognize the possibility. GE has patents on carbon-capture systems using carbonic anhydrase dating to the mid-1960s, and medical researchers have explored its potential for artificial lungs. But those efforts have generally foundered on problems with stabilizing the system—in particular, keeping the liquid membrane from drying out. Trachtenberg has been exploring hydration mechanisms modeled, again, on the body's own strategies, in this case for contending with water loss.

In 1991, Trachtenberg founded a tiny company called Carbozyme, near Princeton, New Jersey, to pursue his research; he has funded it ever since almost entirely on his own, spending several million dollars earned from the sale of a previous start-up (a company that makes "nutriceuticals" to reduce cravings in drug addicts). For many years he worked with the most bare-bones experimental technology: a Bunsen burner to provide the gas stream, and Saran Wrap to seal the joints of his experimental filter.

What he and his colleagues developed is a kind of jelly roll made of a specially engineered membrane (the cake), spread with the catalytic carbonic anhydrase enzyme (the jelly), and filled at the center with a salt solution, all rolled into a cylinder. The membrane is made of tiny polypropylene straws, each the diameter of a toothbrush bristle. Each straw has been stretched during manufacture, like bubblegum, to create nanoscale stress tears big enough to let the carbon dioxide into the jelly roll but not big enough for the enzyme to leak out. The membrane is also hydrophobic—meaning it will not get wet—preventing the water inside the membrane from leaking into the micropores.

As the flue gases are piped past the front of the jelly roll, the carbon dioxide moves into the liquid center of the membrane, driven (as in the body) by differential pressure and reacting (as in our cells) with the enzyme to become bicarbonate stored in the salt solution. On the back side of the jelly roll, the process is reversed. A flow of "sweep gases" (or a simple vacuum) drives the reaction the other way: because those sweep gases are leaner in carbon dioxide than the bicarbonate-stuffed solution inside the membrane, the bicarbonate turns back to carbon dioxide gas, which moves out of the membrane and into a pipe for transport and sequestration. Again, the carbonic anhydrase is key, and not only because it accelerates the carbon dioxide uptake and release. Without the enzyme, every gas in the smokestack mix, from nitrogen to sulfur dioxide, would cross to the lower-pressure "sweep" side of the membrane. But the enzyme acts like the doorman at an exclusive club: catalyzing only carbon dioxide reactions, it quickly sweeps in the carbon dioxide before any other gas molecule has a chance to slip in.

According to Trachtenberg, each filter, which at full scale will be thirty-three inches in diameter and six feet long, will process a ton of carbon dioxide a day from a coal-burning plant. Because each is a self-contained unit, power plant operators will be able to add filters in any number to reach the desired processing volume, arraying them in big flat racks or climbing a winding spiral. Six thousand filters, costing $150 million, would—he says—take the carbon out of the emissions from a 400-megawatt coal-fired plant, using between 10 and 15 percent of the power plant's energy output and pushing the cost of electricity from the plant up 20 percent.

Membranes have been used for many years to clean carbon dioxide from natural gas, but at costs far too high to be practical for the massive volumes of gases that come out of smokestacks. Alstom funds some long-term university research on membranes and enzymes, but Bob Hilton believes the scaling challenges will be immense. "You're talking about millions of cubic feet of gas rushing through a thirty-foot pipe, and they want us to equally distribute it into a couple of thousand three- or four-inch cylinders; that's an incredibly difficult engineering problem." Trachtenberg acknowledges that distributing the gas is a serious challenge, but he believes that very recent advances in computational fluid dynamics—computer simulations of the interaction of fluids and gases with complex surfaces—have made it solvable.

Trachtenberg now has a $7.2 million federal grant, enzyme supply contracts with Novozymes and engineering contracts with Siemens, and is developing a pre-pilot project at the Energy & Environmental Research Center in North Dakota. Over the next several years, he will scale up the system at the center to handle emissions from 1 megawatt of capacity before taking it into the less-controlled environment of a real power plant slipstream. By late 2007, Carbozyme was looking for venture capital. "I've been happy to grow slowly," Trachtenberg says, "but with the competition beginning to heat up, it seems time to move."

ALL OF THE TECHNOLOGIES discussed so far are designed for pulverized-coal plants, and most are intended to be retrofitted on the fast-growing stock of existing plants. But another whole universe of innovation is focused on new plants: turning coal to gas before it is burned, to make pollutant removal easier. The technology has deep roots stretching all the way back to the nineteenth-century "gasworks" that used to rise on the edge of most U.S. towns, producing "town gas" for streetlamps and residential use.

Removing carbon dioxide after coal is burned is hard because of the sheer volume of flue gases to handle, and because those gases are so dilute, which makes necessary the kinds of huge separation systems Eli Gal and Trachtenberg are working to build. Conventional combustion is done with air, which is 78 percent nitrogen, while gasification occurs in pure oxygen and produces only carbon monoxide and hydrogen. That already significantly reduces its volume; because the syngas is also under higher pressure, its volume is reduced 160 times, making the gas far easier to manage. The removal of carbon dioxide from syngas is an established technology: mixed with steam in a "shift reactor," the carbon monoxide converts to carbon dioxide, which can be absorbed with an off-the-shelf physical solvent, leaving pure hydrogen to be burned. Conventional pollutants like sulfur are also more concentrated and thus more easily removed. The hydrogen, when burned, makes carbon-free electricity. In an "integrated gasification combined-cycle" (IGCC) plant, the hot exhaust gases from that hydrogen combustion are used to make steam and power a second turbine, improving efficiency.

Although gasification champions mostly scorn pulverized coal, and though about 160 coal-gasification plants now operate worldwide, debate continues about the relative merits of the two technologies. But this is the beauty of a cap-and-trade system for greenhouse gases: *Under such a policy, it would not matter that no one yet knows which technologies will prove superior.* It would be up to the market—not the government—to find the best ways forward. Instead of forcing utilities to make certain decisions or to back certain technologies, a cap-and-trade system sets a limit and lets the market figure out the cheapest and most efficient means of getting there. Every measure proposed to clean up pollution has a potential role to play. Once a firm upper limit on emissions is set, and a trading mechanism is in place, allowing those who exceed that limit to buy allowances from those who beat their targets, the resulting price signal will determine what cleanup solution makes the most economic sense for any given facility.

GE is betting on IGCC, believing that the 5 to 20 percent higher cost of building such plants will be more than offset by the savings on carbon capture. Phillippe Joubert, president of power systems at Alstom, believes IGCC will be more expensive than pulverized coal, even when carbon capture is factored in. A global carbon market will ultimately sort out which company was right. The $1.5 billion, U.S. government–funded FutureGen project is aiming to build a zero-emissions IGCC plant with sequestration by 2012. And in November 2007, the Indiana Utility Regulatory Commission approved Duke Energy's application to build an IGCC plant, provided the company submits a plan on carbon capture. But IGCC plants in Illinois, Minnesota, Florida, and Colorado have been suspended or canceled, in several cases because the utilities were unwilling to invest in carbon dioxide–capture technology without greater regulatory certainty on both carbon limits and the requirements for underground storage.

Given the absence of a fleet of power plants that can burn syngas, and the lack of a pipeline system for transporting it, a number of companies are taking gasification a further step, converting coal into "substitute natural gas"—nearly pure methane—which can be transported in existing natural gas pipelines and burned in existing gas-fired plants. A Cambridge-based company called GreatPoint Energy has raised $137 million in venture funding to develop catalysts that dramatically speed up the coal-to-methane conversion process—without turning the coal to syngas along the way. GreatPoint says it can produce the methane for $3.50 per million cubic feet, which would revolutionize the marketplace competition between coal and natural gas (in some parts of the United States, recent inflation in coal and coal technology prices has made gas competitive for new power generation at double GreatPoint's price). The company also intends to capture and sequester the carbon released during that process of "methanization," winding up with a product nearly as cheap as coal but twice as clean. Though in the long run, even natural gas will not get us to the 80 percent emissions reductions the United States needs to achieve, the rapid emergence of these technologies will make it possible in the near term for utilities to substitute new natural gas plants for previously planned conventional coal-fired facilities, providing a valuable bridge to still cleaner options. And where carbon dioxide is captured as a by-product of coal gasification, it will also jump start construction of the infrastructure for geosequestration.*

---

* GreatPoint is also developing technology to convert biomass into methane, capturing and sequestering carbon dioxide; because the biomass absorbs carbon dioxide while growing, that could actually subtract carbon from the atmosphere.

Back in Phoenix, Ray Hobbs at APS is running a test reactor that uses hydrogen to "hydrocrack" coal into natural gas, aiming, as Hobbs usually does, to solve several problems at once. First, the storage problem for renewables: wind energy produced at night, when demand for electricity is low, could be used to make the hydrogen. And second, the absence of an infrastructure for the "hydrogen economy": by putting the hydrogen into the hydrocarbon (natural gas), he says, it will become immediately usable.

ONE OF THE MOST RADICAL EFFORTS to remake coal gasification is also one of the oldest, dating to 1868, when Carl Wilhelm Siemens first conceived the idea of gasifying coal right where it is found—underground. Siemens proposed drilling a hole deep into a coal seam, igniting a fire, and letting the heat produced by the burning edge of coal do the work of gasifying the rest, which could then be piped to the surface. The coal seam itself, in other words, would become the gasworks.

The development of underground coal gasification (UCG) would stop and start, sometimes violently, many times over the next century, a reminder of the huge historical forces that shape energy, and of energy's shaping force on history. Dmitri Mendeleyev, the Russian chemist credited with conceiving the periodic table of the elements, picked up Siemens' idea and worked on it for several decades; by 1913, British scientist Sir William Ramsay (who had won the Nobel Prize for chemistry in 1904 for his discovery of the noble gases) was raising money from industrial investors to build the first underground coal gasification plant. World War I shut down that effort, but Ramsay's speeches had attracted the attention of a Russian radical then in exile in Zurich, V. I. Lenin. In 1913, Lenin published an article in *Pravda*, celebrating the promise of this "Great Victory of Technology" to liberate workers from the brutal and dangerous work in the mines. Fifteen years later, Stalin took up the cause, both to demonstrate his own Communist fealty to the workers and to ratchet up the competition with Germany, then making rapid progress on Fischer-Tropsch technology. The Soviet leader put thousands to work on it. In 1937, when the first team failed to reach its targets, many were put on trial and several were shot.

All told, historians estimate, the Soviets invested $10 billion (in current dollars) in the technology. By 1939 they had successfully started operations at a plant in Ukraine, but were again shut down by war and German occupation. Restarting the program after the war, they got fourteen plants up and running, producing fuel for heating and cooking, before Krushchev shut most of them down. He wanted to erase all remaining vestiges of Stalin. And the country had discovered huge reserves of natural gas.

Just two Soviet underground coal gasification plants remained in operation, one in Siberia and one in Angren, Uzbekistan. It was in Uzbekistan that Michael Blinderman and Simon Maev, the two men now driving the resurgent development of this technology, first met, in 1971, in the ninth grade at a Tashkent high school for physics and math. Maev, the blue-eyed charmer of the pair, had grown up at the Uzbek plant, which his father ran until 1965; Blinderman, stern and serious, went to Moscow to study physics and wound up overseeing research and development for the entire Soviet underground coal gasification program. Though the Soviets made great advances in relevant technologies, including directional drilling, they did little to bring down costs, since central planners owned both the power plant and the gas reactors and set the terms of sale between them.

In the midst of the 1970s energy crisis, the United States began its own experiments. First, the Williams Brothers Engineering Company of Tulsa, Oklahoma, purchased licenses from the USSR—the single biggest technology transfer ever from the Soviet Union to the United States—and launched several trials in Texas. Though the company abandoned those experiments when the price of natural gas fell, over the next two decades some thirty more research projects in underground gasification were conducted, many of them overseen by the Lawrence Livermore National Laboratory. Two ended badly. At one, the lab lost control of the underground fire. At Hoe Creek near Gillette, Wyoming, operators overpressurized the cavity and drove a plume of pollutants into three freshwater aquifers, requiring cleanup of carcinogenic benzenes and phenols that continues to this day.

The attempt to commercialize the technology in North America, however, did not begin until the early 1990s, when Blinderman, Maev, and a handful of their colleagues from the old Soviet program emigrated to Montreal and founded Ergo Exergy. By 1997 they had their first demonstration project operating in Chinchilla, Australia, 300 kilometers west of Brisbane, gasifying a coal seam 900 meters deep, but they shut it down when they failed to secure ongoing financing. Chinchilla now has new capital and may restart in 2008.

By 2007, Ergo Exergy had its first commercial project operating in South Africa, which makes three-quarters of its electricity from coal and made deep investments in Fischer-Tropsch coal gasification technology during the apartheid years, when international sanctions shut down energy imports. At its 4,200-megawatt Majuba power station outside Johannesburg, Eskom, the South African state utility, is mixing Ergo Exergy's syngas into conventional boilers, replacing about 30 percent of the pulverized coal. Although the plant is situated right at the mouth of a mine, unsafe geological conditions had shut down mining operations and forced Eskom to truck in coal more than a hundred miles; with underground gasification, they have been able to resume use of the local resource. Eskom is also building a 2,100-megawatt combined-cycle plant on the same spot to burn the Ergo Exergy gas.

Underground gasification is now in development all over the world. China's program is the largest (though Blinderman will not work there, out of concerns about intellectual property). In 2007, India's national utility company began negotiations with Blinderman; BP licensed the Ergo Exergy technology for development in Wyoming, and GE and Siemens both warranted their turbines for use with Ergo Exergy's gas. Laurus, the company founded by Simon Maev that holds the Canadian license for the technology, was pressing toward commercial production by 2011. Maev's ambition over the next decade is to shift all of Canada's twenty coal-fired power plants to underground coal gasification technology, first by mixing gas into the boilers as South Africa does, then by replacing plants, as they retire, with combined-cycle facilities. If he succeeds, Maev says, Laurus all by itself will enable Canada to meet the carbon-reduction targets mandated by the Kyoto Protocol on climate change—if they can find a way to sequester the carbon dioxide.

Here's how the technology works. First, the operators drill two holes, a few hundred feet apart, deep into the coal, beginning in the portion of the seam that is closest to the power plant. Then they light a fire in the first hole while also injecting air, oxygen, or steam, depending on the end product they want to produce. As the coal begins to burn, it turns the solid wall of coal in front of it into gas: That hot gas (about 1,200°C, or 2,200°F) acts as a cutting tool, burrowing a tunnel through the seam (the coal is the only combustible substance, so the burn naturally follows the seam). The reactor, in effect, is the earth itself. The operators control the process by adjusting the pressure and composition of the injection gases and the rate at which they take the gas out the other end.

ADVOCATES OF UNDERGROUND GASIFICATION claim many advantages, both environmental and economic, over pulverized coal and surface gasification. To begin, there is no mining: no

Appalachian mountaintops stripped bare; no open pits, tailings, ash dams, polluted runoff, coal crushing, trains to transport the coal; no lives lost to underground floods, explosions, fires, or pneumoconiosis. Second, underground gasification makes accessible vast energy resources now out of reach. In the United States alone, according to a report from the Lawrence Livermore National Laboratory, the technology would increase recoverable coal reserves by at least 300 percent.★ Coal gas made underground is less costly to produce, saving not only on the capital and operating expenses of mining, including the huge energy expenditures for mining and moving the coal and the waste, but also on the cost of building a gasifier. The bottom line is cheaper power. At its South Africa operation, Ergo Exergy produces its gas for $1 per million BTUs, one-third to one-sixth the cost of gas made in a surface gasifier.

Most important for the atmosphere, the net emissions of carbon dioxide and other pollutants from underground gasification may prove to be lower than those of any other coal technology. Because temperatures are lower in underground gasifiers, and because the coal reacts with some of the water inside the coal seam, Ergo Exergy winds up with a different mix of gases at the end. Half the carbon in the coal comes up right away as carbon dioxide, ready to be captured with off-the-shelf absorbents like Selexol, at a cost of about $25 per ton. A number of pollutants, including mercury, sulfur dioxide, nitrogen oxides, particulates, and ash, are for the most part left in the ground, according to the Lawrence Livermore lab report. What remains is the fuel: a mixture of hydrogen and hydrocarbons that, when burned, will have a carbon footprint equal, according to the Lawrence Livermore scientists, to a natural gas combined-cycle plant, the cleanest of

---

★ "Best Practices in Underground Coal Gasification," 2007.

all fossil fuel power plants.★ To get to near-zero carbon emissions, the company could go a step further: adding a shift reactor to remove the remaining carbon from the syngas, and sequestering that as well. Combining the cheap cost of making the gas with the cheap cost of cleaning it, says Blinderman, they will be able to make electricity with carbon capture and sequestration for $30 a megawatt-hour, less than half the cost of an IGCC plant with carbon capture, and not much above the cost of pulverized coal today *without* carbon capture.

The two biggest environmental concerns remaining, in the view of the scientists at the Lawrence Livermore lab, are the potential for groundwater contamination and for surface subsidence as the underground cavity collapses. Both, they conclude, are manageable with good site selection (away from groundwater resources, within good geologic seals) and the use of "best practices" in their operation, including advanced computer simulation. "Ergo Exergy's recent pilot at Chinchilla, Australia," the report concludes, "demonstrates that it is possible to operate [underground coal gasification] without either hazard."

Michael Blinderman explains that the company always works below the water table and at pressures low enough so that water will only flow down into the coal seam, not out or up. Ergo Exergy continuously monitors those pressures, as well as the composition of the groundwater; the Chinchilla facility had nineteen monitoring wells. The company claims, in fact, to be a net producer of clean water.

The roof of the underground cavity carved by the fire does sag and deform, Blinderman concedes, collapsing by up to 30 percent. In a hundred-foot-thick seam of coal, that could produce a thirty-

---

★ A fuel's carbon dioxide "cleanliness" is determined by its carbon-to-hydrogen ratio. Coal is the dirtiest, with a ratio close to 1:1. Oil's ratio is roughly 1:2. And methane, the main component of natural gas, has a ratio of 1:4. Pure hydrogen, which contains no carbon at all, is the cleanest.

five-foot depression, altering wells or aquifers beyond use. At Chinchilla, as gasifiers advanced along the coal seam, capping finished
wells behind them, farmers followed close behind, quickly restoring the land to agricultural use. The Lawrence Livermore scientists
note, however, that such a small-scale project does "not accurately
represent commercial large-scale evacuation" of coal.

A last intriguing possibility held out by Blinderman and company is to use the underground cavity created by the fire to store
the carbon dioxide that is captured from the gas, compressing and
reinjecting it on the spot. The cavity would be big enough to hold
all the carbon dioxide captured on site, they say. And they will have
already been assured that it has good caprock, since they would
have lost all their commercial gas if the cavity leaked. They are
now developing a New Zealand pilot to demonstrate such "reactor zone storage."

The idea has promise, according to the national lab, because coal
"swells and plasticizes" in the presence of carbon dioxide, "closing
fractures and porosity fairly quickly," and because the carbon dioxide will likely, over time, be adsorbed onto the mineral surfaces and
immobilized. But the scientists also think the problems with reactor zone storage may take a decade or more to solve—if, indeed,
they can be solved. For instance, the carbon dioxide might interact
with the water and char left after the coal has been burned, to
produce carbonic and sulfuric acids that would leach harmful metals. Or the rock may simply have been too stressed by temperature
fluctuations and subsidence to be reliable.

In fact, every carbon-capture technology discussed in this chapter depends on a big challenge: Can geologic sequestration work?
Questions remain about precisely how to ensure that carbon
pumped into the ground will stay there; what collateral impacts
large-scale sequestration might have, and how much the pipelines,

compressors, and monitoring for the whole operation will cost.★
Bob Hilton of Alstom points to the complex property-rights issues
raised, since the gas can spread out as far as thirty miles underground. The infrastructure challenges will be equally daunting.
Charles Christopher, a sequestration consultant to BP in Houston, calculates that sequestration sufficient to achieve just 15 percent of the necessary global reductions in carbon dioxide would
require handling one and a half times the daily volume of material
handled by the entire world oil industry. If just two-thirds of the
total amount of coal emissions in the United States were liquefied
for storage, that liquid carbon dioxide would fill as many barrels as
hold all the oil the country consumes. And without a legal limit on
dumping carbon into the atmosphere, and therefore no monetary
value attached to that carbon, who is going to pay for all that infrastructure and handling?

The technology of sequestration is developing rapidly. Every
year for a decade, the Norwegian state oil company, motivated by
Norway's carbon tax on offshore oil and gas (about $80 a ton in
December 2007), has been injecting a million tons of carbon dioxide deep into the bed of the North Sea. At a natural gas field in
In Salah, Algeria, BP and the Algerian state-owned energy company are injecting another million tons a year deep into the desert. And at the Weyburn oil field in Saskatchewan, Canada, 8,000
tons of carbon dioxide a day, about equivalent to the emissions
of a 400-megawatt coal plant, is piped in from the Great Plains
Synfuels Plant in North Dakota and used to increase the pressure
in a partly depleted oil field. Continuous monitoring at all three
sites has detected minimal leakage. In 2005 the Intergovernmental

---

★ In October 2007 the EPA committed to writing rules, under the Safe Drinking
Water Act, that "ensure there is a consistent and effective permit system" for carbon
sequestration on a commercial scale.

Panel on Climate Change concluded that with appropriate selection and management of geological reservoirs, retention rates are likely to exceed 99 percent over a thousand years.

A major informal test of carbon sequestration has been underway for several decades at the more than seventy enhanced oil recovery projects operating around the world, though these sites were not designed for carbon dioxide storage and are unmonitored, so it is unknown how much of the gas is remaining underground, or for how long. The greatest value in these enhanced oil recovery projects has been in the financing they have drawn into carbon dioxide pipelines and other infrastructure.

Sally Benson, executive director of the Global Climate and Energy Project at Stanford University, believes there is sufficient capacity in the world's deep salt basins, mature oil fields, and (possibly) deep coal seams to hold several centuries' worth of emissions. The briny, porous sandstones at the bottom of the North Sea and Gulf of Mexico, many of them capped by impermeable shales, are particularly promising. The rock traps carbon dioxide in its pores, which means that to go anywhere, the gas must move from pore to pore. Over decades, as it dissolves, carbon dioxide loses its buoyancy, becoming as dense as oil and forming connected bubbles held together by capillary forces, further immobilizing it. Natural gas has more buoyancy than carbon dioxide, Benson notes, but has been stored in underground reservoirs for tens of millions of years. Over centuries, the carbon dioxide will react with other minerals to form carbonates and become part of the solid rock matrix.

Not surprisingly, some of the entrepreneurs invested in renewable energy remain skeptical about sequestration. John O'Donnell of Ausra (the solar thermal company discussed in Chapter 3), riffing on Oklahoma Senator James Inhofe's infamous dismissal of climate change, calls carbon capture and sequestration "the greatest hoax ever perpetrated on the environmental community."

Clean-coal advocates, of course, take a more positive view. Simon Maev believes Ergo Exergy and Laurus will be particularly well equipped to find geological "sinks" in which to store the carbon dioxide, even if their reactor cavities do not prove viable. "We already know the hydrogeology of the area, the rock mechanics," says Maev. "The exploratory drilling to find those sinks costs up to $50 million, and we're already partway there."

Eli Gal is confident sequestration will work long enough to matter. "We don't need solutions forever," he says. "We're in a transition. Fossil fuels are finite, and we'll have to move in the next decades to renewables anyway. So the problem is the next twenty years." For Eric Redman, there is simply no other choice. "There is no possibility of stabilizing the climate without massive sequestration. So to stand here arguing about it without moving forward is like standing on a sinking ship and hesitating to get in a lifeboat because you worry it might leak. This lifeboat has to get us to shore."

Today's Solutions

On a sweltering day in December 2002, in Pará state in the Brazilian Amazon, Herculano Porto thought the world was coming to an end. Machete in hand, he was clearing leaf litter and brush from the forest floor beneath his 150-foot-tall Brazil nut trees, to make it easier in the coming months to find the coconut-sized outer shells that hold the nuts when they fall. He had paddled from home across the Anfrízio River to work in the grove where he had buried his father years before. Then he saw four men with guns coming through the forest.

It would be two years before those gunmen would leave Herculano's forests, two years that would take Herculano on an unimaginable journey, including to the office of Brazil's president. Herculano cannot read or write. He is not an inventor or a venture capitalist, not an engineer or biochemist or CEO. But he and his rainforest neighbors may be the first to pull the world back from catastrophe. Herculano would wind up playing a part in a policy innovation as transformative as any technological breakthrough yet dreamed of: by changing the rules so that illegal destruction of forests ceased to be profitable, he and his allies would stop that destruction—at least for a time. In so doing, they would also create a template for a global reform of immense possibility, a reform

now at the center of international climate negotiations designing the follow-on to the Kyoto Protocol. Ultimately, this reform could stop deforestation for good.

Herculano also has in his forest backyard a shortcut to safety: one of the quickest, easiest, cheapest ways to grab huge and immediate carbon reductions. Until now, this book has focused on possibilities for reducing global warming pollution that are still just out of reach. But this chapter considers the many opportunities already lying at our feet, waiting to be scooped up like Herculano's Brazil nuts as soon as there are profits to be made. Simply using energy more efficiently, for example, can make an enormous difference, easing the transition to a low-carbon economy by leveraging known technologies we can rely on now—while other solutions incubate.

SLIGHT AND SOFT-SPOKEN, with a creased face and graying hair, Herculano is one of the *ribeirinhos*, or river-folk, who live scattered across the vast Amazonian interior, an area as large as the United States west of the Mississippi. His story comes to us thanks to Steve Schwartzman, an anthropologist on staff at Environmental Defense Fund who has worked in the Amazon for twenty-five years. Steve met Herculano in the Brazilian port of Altamira in 2003 and has visited his home on the Anfrízio River.★

Herculano's people live mostly on game, fish, and manioc, the starchy root crop they grow in their gardens, along with the little bit of money and bartered goods they make from gathering forest nuts, resins, and oils. When he was younger, Herculano spent long hours hiking the jungle trails in search of rubber trees, tapping the

---

★ Since then, Environmental Defense Fund has coordinated a project with Brazilian nongovernmental organizations, including the Instituto Socioambiental and Fundação Viver Produzier e Preservar, to protect the land of Herculano and his neighbors and to improve their living standards.

milky latex and making it into fifty-pound dry balls over a smoky fire. It was back-breaking, suffocating work, and he was at the mercy of the estate bosses who controlled the nut and rubber trades and paid what they liked. For a kilo of Brazil nuts they might pay 50 cents; for twelve kilos (twenty-six pounds), Herculano might get a can of condensed milk for his children. Still, he remembers those days wistfully: the school, the boat that twice a month steamed for several days to Altamira, 250 miles away, returning with salt, sugar, and coffee.

Hard as it had been, Herculano's life grew much harder in the 1970s and 1980s, when the price of rubber collapsed and Brazil's military government launched its massive effort to build and pave roads to open the "Amazonian frontier." The rubber bosses left, closing the school and clinic behind them. In their place, eventually, came the *grileiros*, the land-grabbers, who as soon as a patch of forest was made accessible would arrive to log or burn it down, then sell the cleared land to ranchers. By 2001, when the road finally reached Herculano's forest and the gangs began offering to pay him and his neighbors to first clear and then evacuate their land, some people were so poor and scared that they took the bit of money the speculators offered and moved to Altamira or beyond. Herculano saw the crude "no trespassing" signs the gunmen were hammering onto trees, knew they were telling him he couldn't hunt anymore or collect nuts or plant gardens. But he had also been to the city and seen the shantytowns where men like him struggled to keep their children fed and out of trouble. So he turned their money down.

Then the outsiders started saying that other people had already bought the land—important people with money and connections, like Osmar Ferreira, the logger up in Trairão on the Transamazon highway, or Mr. Julio, a judge with a big cattle ranch in Mato Grosso. It was Mr. Julio, they told Herculano on that frightening day in 2002, who now owned the Brazil nut grove he and his father had harvested for six decades, and had given orders to "put the forest on the ground" before the rains were done. Herculano didn't challenge the men. He just gathered his things and paddled back across the river to his home. He was scared and couldn't think what else to do. Gradually, all around him, the forest began to fall.

FOR NEARLY THREE DECADES NOW, similar scenes have been playing out across the Amazon, destroying some 5 million acres of rainforest each year. Business as usual, some models predict, could turn half the Amazon forest to savannah by 2050.

Worldwide, driven by the profits to be made from tropical hardwoods, soybeans, and beef, between 20 and 40 million acres are destroyed each year. When standing, these forests are like great vaults, locking up billions of tons of carbon in their root structures and wood. Bringing them down is like opening Pandora's box— releasing all manner of misery upon the world. Tropical deforestation wipes out the livelihoods and lifeways of innumerable poor, marginalized people like Herculano. It erases the world's most diverse ecosystems, jeopardizing the continued existence of creatures like the giant otter, whitelipped peccary, and hyacinth macaw. And it contributes fully one-fifth of all global warming emissions, as much as the whole U.S. economy, or every car, truck, and bus in the world combined.

Tropical deforestation is in fact the second-biggest cause of global climate change, trailing only the burning of fossil fuels. Climate change, in turn, accelerates the forests' demise and the impoverishment of the people who live there. Droughts are already intensifying in Indonesia and the Amazon, killing animals and fish, compromising freshwater supplies and increasing the frequency and intensity of wildfires. Climate change and poverty are deeply entwined.

Just as hope remained with Pandora, however, hope arrived for

Herculano just a few months after that terrifying day with the armed men. A new kind of visitor began showing up at his door, telling him of a way he might protect his land and continue supporting his family as he always had. A young man named Tarcísio came from the Roman Catholic Prelacy of Xingu, in Altamira, bringing with him a local leader named Antonia Melo. They spoke of an idea first championed by Chico Mendes, the legendary rubber tapper who had organized his fellow workers and fought to defend the rainforest until he was murdered by cattle ranchers in 1988. Mendes had proposed the creation of a new kind of protected area called an "extractive reserve," which would allow people like Herculano to continue their fishing and gardening but would prohibit activities that harmed the forest. If Herculano would sign a letter supporting such a reserve, Tarcísio and Antonia told him, and get his neighbors to sign as well, the federal government would come protect their land and provide community health services and education. Herculano wasn't sure he believed it all, but he started talking to his neighbors, persuading most of them to sign.

Then the death threats began.

Raimundo Pereira, a gang leader, told a friend of Herculano's to pass on a warning: "The first person who brings the feds here will be the first one to die." The threats were all too credible. Since 1971, more than seven hundred rural workers and local leaders had been murdered in Pará. As Antonia and Tarcísio worked to push establishment of the reserves through the bureaucracy, they grew increasingly worried; Environment Minister Marina Silva was taking too long to sign the decree creating the Anfrízio reserve. They begged their contacts in the ministry to act quickly before Herculano was killed. In October 2004, Herculano's allies in Altamira sent a helicopter into the jungle to get him out; with Tarcísio, he then flew to Brasília to make a direct appeal to President Luis Ignácio Lula da Silva. A month later, the president signed the decree protecting the entire Anfrízio watershed.

The creation of the reserve had an immediate impact. Until then, the government viewed the Anfrízio region as "empty land" (*terra devoluta*). Though publicly owned, it was essentially ungoverned: anyone providing any evidence of occupation could claim title to great chunks of land. Since deforestation was considered evidence of occupation, *grileiros* had every incentive to clear as much land as they could, then sell it and move on to the next new frontier.

Once the land was enclosed in an extractive reserve, however, speculators could no longer claim title, nor could they convince buyers that they ever would. Paying hired guns to clear forest and chase off river people from land that they could never claim or sell was an obviously poor investment. Over the next few months, the gangs gave up and disappeared. Tarcísio and Antonia got Herculano and his neighbors a boat to carry goods to market, and a shortwave radio so they could keep in touch with neighboring communities and their allies in Altamira. In February 2005, when the government sent six soldiers to keep the peace in Anfrízio, there wasn't much for them to do but listen to the radio and fish. They even taught some kids their ABCs.

EXTRACTIVE RESERVES, indigenous lands, and parks now cover nearly 40 percent of the Amazon—733,559 square miles, an area almost three times the size of Texas. But that leaves an equal area of the Amazon with no legal protection. And even within the reserves, the pressures remain fierce. As soy prices increase, so do the fires. The murders continue: In 2005, the killing of Sister Dorothy Stang—a Roman Catholic nun from the United States who had spent thirty years fighting against the destruction of the rainforest and its people—made clear the lengths to which those determined to profit from deforestation were still willing to go. In response to that killing, and at the urging of Herculano's allies in Altamira and Brasília, Brazil's government designated new reserves that connected two existing indigenous territories, thus creating a 69-million-acre continuous corridor of protected areas. Called the Xingu Protected Areas Corridor, it is the largest mosaic of protected areas in the world.

For a few years, deforestation declined. By 2007, however, sporadic law enforcement, lagging efforts to remove invaders, and rising commodity prices conspired to cause it to resume. The reserves had removed one of the chief incentives for destruction. But as yet, incentives for protection remain missing.

Herculano's allies have therefore also been working on a larger stage: the international negotiations designing the post-Kyoto global regime. The core insight remains unchanged. The rainforest will be used to the most profitable end, whatever that end might be. The only sure way to preserve it, therefore, is to make it worth more to people intact than destroyed.

In September 2007 three major Amazonian grassroots organizations held the Second National Meeting of the Peoples of the Forest in Brasília. The first meeting, held just months after the assassination of Chico Mendes in Rio Branco, capital of Mendes's home state of Acre, had brought together a few hundred Indians and rubber tappers. The second drew two thousand delegates and participants, including some of Herculano's Anfrízio neighbors, cabinet ministers, and the president.

One of the principal subjects of discussion was an idea first proposed in 2003 by a group of researchers from the Instituto Socioambiental, Amazon Institute for Environmental Research, Environmental Defense Fund, and others, and put on the official UN climate agenda in 2005. To protect the rainforest, achieve the largest-scale carbon reductions possible in the near term, and create a path for developing nations into the global emissions reduction regime, tropical countries should earn tradable carbon credits for reducing deforestation on a national scale. Two and a half acres of rainforest contains between 120 and 300 tons of carbon; preserving half a million acres of that forest would keep out of the atmosphere about as much global warming pollution as a 500-megawatt coal plant emits in its entire fifty-year lifetime. The inclusion of "forest protection credits" would also help overcome one of the greatest obstacles in global climate negotiations: how to engage key developing countries, particularly Brazil and Indonesia, in reducing emissions without choking off modernization and economic growth. (Because of deforestation, Indonesia and Brazil rank third and fourth in the world—behind China and the United States—in greenhouse gas emissions.)

Here is how carbon-credit trading would work: Brazil and the other participants in the global carbon market would agree to use a recent historic period as a baseline for comparison. Brazil would then have to reduce the rate of deforestation below that baseline. (Brazil has accurately monitored and measured its forests via satellite since the 1970s, so if it succeeded in reducing national deforestation it would be clear in the satellite data.) A ton of carbon not emitted could be sold in the global carbon market to companies needing additional reductions. For example, a German power company that emitted 1.2 million tons of carbon dioxide, but only had allowances for 1 million tons, would need to buy 200,000 tons of reductions; one of its options in a global carbon market would be to pay Brazil for the carbon credits it had earned by reducing deforestation ($6.3 million, at November 2007 prices). The total value of allowances traded in carbon markets worldwide (created in preparation for the first Kyoto Protocol commitment period, which begins in 2008) reached $30 billion in 2006. Eventually, the carbon market is expected to be worth hundreds of billions of dollars annually. The potential of claiming even a small fraction of this market would be a powerful incentive to preserve forests.

For Brazil, getting paid to keep forests standing would generate more profits per acre than most of the cattle and illegal logging operations, and create a stream of revenue for financing conservation and protecting forests far beyond what Brazil's government and international aid could ever supply. In late 2007, Daniel Nepstad of the Woods Hole Research Center and colleagues determined that over the next thirty years, Brazil could prevent emissions of 6 billion tons of carbon at a cost less than $2 a ton. In a market that priced carbon at $10 a ton, those avoided tons would net Brazil $48 billion in profits; at $30 a ton, Brazil would earn $168 billion.

In his opening statement at the September meeting, indigenous leader Jecinaldo Sataré Mawé urged President Lula to move Brazil to the fore of the fight against climate change by advocating for "just compensation for reduced deforestation [through the] financial mechanisms of the carbon market."

Creating a market for such carbon credits will again depend on Herculano and his neighbors. The most extensive and durable reserves on Brazil's Amazon frontier are not national parks, which are few, but the inhabited indigenous and extractive reserves, where there are people on the ground with a direct stake in the forest's survival. A market for carbon, or anything else, cannot exist in a lawless place where everything is up for grabs, and it is people like Herculano who are forcing officials in Brasília to sort out functional governance for the Amazon.

When there are profits to be made by keeping these forests standing, the same entrepreneurial energy that has brought them down at such an alarming rate will become an equally powerful force for their protection.

THE TROPICAL RAINFOREST may seem a universe away from the insulation in your attic or the car in your garage. But just as immediate and immense carbon reductions can be achieved by reducing deforestation, opportunities far closer to home offer abundant low-hanging fruit awaiting harvest.

One of the most important lies in eliminating other greenhouse gases. Methane, for instance, is the second most common human-produced greenhouse gas after carbon dioxide and has a heat-trapping effect more than twenty times that of carbon dioxide.* But methane also has something carbon dioxide does not: a positive value as a fuel. Increasing numbers of farmers and landfill managers have therefore begun capturing the methane gas emitted by animal wastes and trash dumps to produce clean, cheap power. Burning the methane eliminates most of its global warming impact.

In 2007, 125 biogas recovery systems in the United States were reducing methane emissions from manure by about 80,000 metric tons—with a warming impact equivalent to 1.7 million tons of carbon dioxide—and generating 275 million kilowatt-hours of energy. The EPA says about seven thousand farms could potentially make a profit by turning their manure into gas, providing 700 megawatts of energy to rural areas while avoiding 1.3 million metric tons of methane emissions.

At the Whitesides Dairy in Idaho's Magic Valley, Steve and Brent Whitesides scrape manure produced by their six thousand dairy cows into covered tanks owned by Intrepid Technology and Resources, where it ferments without oxygen into gas. Cleaned of carbon diox-

* HCFCs—hydrochlorofluorocarbons—the chemicals that destroy the ozone layer, are also a potent greenhouse gas. In September 2007, the 191 countries that signed the 1989 Montreal Protocol agreed to speed up by ten years the phaseout of these damaging chemicals, which will reduce greenhouse gas emissions by as much as 25 billion tons of carbon dioxide equivalent.

ide, hydrogen sulfides, and water (along with nitrogen, phosphorus, and potassium),* the gas is then delivered directly to Intermountain Gas Company. In its first year of operation, the dairy processed 10 million gallons of manure, producing enough pipeline-quality natural gas to heat five thousand homes. Intrepid expects to recoup its $5 million investment in three years; in the future it expects to earn additional revenue by selling both carbon credits and the dried, leftover digested fiber, which makes a gardening product similar to peat moss.

Some cattle and pig farmers use the biogas they capture from fermenting manure to generate electricity on site, for use on the farm or sale into the grid. Researchers at North Carolina State University are exploring a further possibility: using high-temperature gasifiers to turn the solid residues to fuel, which could not only increase energy production but also solve the mounting pollution problems associated with livestock production and animal wastes.

S.C. Johnson & Son's cleaning products plant in Sturtevant, Wisconsin, where Windex and other household cleaners are made, uses methane gas captured and piped in from a neighboring landfill in its cogeneration power plant. (A cogeneration plant makes both electricity and heat.) When the landfill gas-powered turbine was installed in 2003, it cut the plant's carbon dioxide emissions in half—the equivalent of taking more than ten thousand cars off the road.

Landfills are the largest human-related source of methane in the United States, accounting for a quarter of all methane emissions. In 2006, about 425 landfill gas energy projects captured the warming equivalent of 89 million tons of carbon dioxide, while generating about 10 billion kilowatt-hours of electricity (some of the gas was burned for heating). The EPA has identified an additional 560 landfills that could profitably be tapped. In September

---

* The nitrogen and potassium—both needed in Idaho soils—are spread on nearby farm fields. The phosphorus binds to the fibers the dairy will sell to gardeners.

2007, Houston-based Waste Management announced that it would develop methane energy projects at sixty of these sites, adding to the one hundred it already operates.

PERHAPS THE SINGLE MOST IMPORTANT immediate opportunity to reduce greenhouse gas emissions—the nation's first line of defense—is energy efficiency. A November 2007 report published by McKinsey & Company found that the United States could cut projected greenhouse gas emissions almost in half by 2030 without major new technology or lifestyle changes, provided we have the right policy incentives and start soon.★ Fully 40 percent of those reductions would more than pay for themselves, creating net savings. Efficiency would provide about a third of the total. Cars present another big opportunity, which will be discussed shortly.

California has provided a model. According to the state public utility commission, its three decades of efficiency programs have saved 40,000 gigawatt-hours of electricity and eliminated 12,000 megawatts of peak demand—which means that two-dozen 500-megawatt power plants did not have to be built. Even with the most efficient natural gas technology, those plants would now be dumping 15 million tons of carbon dioxide into the atmosphere each year (if conventional, subcritical coal-fired plants, they would be dumping nearly 40 million tons a year). The state's utilities have played a role in reducing demand. Pacific Gas and Electric Company, for instance, has enrolled hundreds of business customers in "demand response programs": when electricity demand threatens to exceed supply, the utility can remotely dim its customers' lights or cycle their air-conditioners on and off to avoid having to fire up another power plant.

★ "Reducing U.S. Greenhouse Gases: How Much at What Cost?" Support for the study came from Environmental Defense Fund, Natural Resources Defense Council, PG&E, Shell, and others.

Entrepreneurs producing software to measure and manage "negawatts," or negative watts, have collected tens of millions of dollars in venture capital in the past few years. One company, ConsumerPowerline, contracts with big energy consumers—chain stores, hospitals, and commercial buildings—to reduce electricity use on short notice when spikes in demand threaten a power outage. At three Sears and six Kmart stores in Connecticut, for instance, the company has installed software to control lights and air-conditioners; when the New England independent system operators issue a "demand response notification," ConsumerPowerline can cut energy use at the nine stores by 540 kilowatts within thirty minutes. The utility pays for those 540 kilowatts, and the company shares the money with the stores. The entire customer base benefits: while a utility pays up to $2 million for the capacity to generate a new megawatt at a power plant, it typically pays less than $100,000 for each megawatt a demand response company can commit to eliminating from the load.

Sometimes called the "first fuel" (before coal, gas, and nuclear), efficiency is without question the cleanest, cheapest energy resource we have. While it costs 6 to 8 cents to generate a kilowatt-hour from new coal or nuclear plants, plus another 2 to 4 cents to move that power through the grid, *saving* that same kilowatt-hour costs just 3 to 4 cents.

The list of energy-saving opportunities grows longer by the day. Cities replacing incandescent bulbs with light-emitting diodes have cut energy use in traffic signals by 92 percent. By installing skylights, sensors to dim in-store lights, doors on its refrigerated display cases, and dozens of other seemingly small improvements, Wal-Mart has cut energy use by 20 percent in existing stores, 50 percent at a few demonstration stores. In May 2006 the company installed auxiliary power units—small, efficient diesel engines—on

all Wal-Mart trucks that make overnight trips. Rather than letting the big truck engines idle during breaks, drivers can turn them off and use the little auxiliary unit to warm or cool the cabin and run communication systems. In a single year, that one change saved Wal-Mart $25 million and eliminated 100,000 metric tons of carbon dioxide emissions.

But the opportunities for innovation are far from spent, and dozens of start-ups are devising new ways to save energy.

Some are focused on individual products. A Silicon Valley company called Spudnik, for instance, is making energy-efficient flat-screen TVs for environmentally conscious couch potatoes. Minnesota-based SAGE Electrochromics makes windows that change color in intense sunlight, without losing transparency, to keep heat out and reduce air-conditioning loads.

Others are devising ways to make efficiency more automatic, without the individual, daily effort that often gets lost in the face of life's many distractions. A Seattle company called Verdiem, launched in 2001, makes software that allows schools, businesses, and government agencies to power down idle computers throughout their networks. By the summer of 2008 it projects that its customers will have cumulatively saved over 482 million kilowatt-hours of electricity and more than $48 million, achieving a reduction in greenhouse gas emissions equal to conserving 31 million gallons of gasoline.

Still others are plugging the big energy leaks built into existing systems. Nextek Power Systems on Long Island makes a device that connects renewable energy sources that generate DC power directly with electronic devices and data centers that use DC, avoiding the energy losses in converting into and out of AC. Google and Intel are leading a coalition working to replace current computer power supplies, which lose about 50 percent of incoming energy, with a new 90 percent efficient global standard. That switch, they calculate, would save $5.5 billion worth of energy a year.

The biggest arena of innovation may be "energy intelligence," which in 2006 attracted more than $450 million in venture capital. The idea is to build the energy equivalent of the Internet: a sophisticated web that draws electricity from where it is abundant and sends it to where it is needed; such a system reduces the need to add new plants in order to meet peak demand requirements. Some of the companies involved are focused on the supply side: developing systems to manage the emerging multidirectional network of distributed power-generating facilities (including intermittent generators, like rooftop solar arrays). Others are focused on the demand side: adjusting consumption levels minute by minute in response to dips in supplies or spikes in price.

GridPoint, a company based in Washington, D.C., combines the two. It has developed a refrigerator-sized device that is essentially a very smart battery, which utilities install in customers' basements. At times of excess supply, coming either from centralized power plants or from renewable systems, the system puts energy into the battery; when demand climbs or generation falls off (the wind stops blowing; the sun doesn't shine), it takes energy back out again. GridPoint also provides the utility and homeowners with control systems, enabling them to remotely adjust thermostats and appliances.

When he founded GridPoint in 2003, CEO Peter Corsell called it "TiVo for electricity." As the company's ambitions and financing have grown (it had raised $88 million by late 2007), Corsell has renamed it a "virtual peaking power plant"—that is, a way to meet peak demand with advanced information technology rather than new boilers and turbines. The system is also designed to accommodate new technologies as they emerge, including plug-in hybrid cars (see below) and fuel cells.

What follows is a very selective sampling of a few more cutting-edge efficiency innovations.

PAX Scientific: Growing up on the beaches of Australia, Jay Harman noticed that seaweed ripped easily in his hands but held together in the roughest surf. It survived those wrenching forces, he learned, by twisting itself into a coil to allow water to pass through with the least possible resistance. Working as a naturalist with the Australian Department of Fisheries and Wildlife, Harman began a rigorous study of the vortices he realized were all around: mapping the flow patterns of ocean and air currents, of tidal eddies and spiraling storm clouds. Harman was not the first to notice nature's preference for a swirl; Leonardo da Vinci spent the last ten years of his life painting whirlpools. But Harman may be the first to have turned his insights into the energy saved by swirling fluids into a series of successful companies. In 1982 he founded ERG, which made products like afterburners for aircraft engines (which inject additional fuel into the jet pipe downstream of—"after"—the turbine, thus increasing engine thrust) and became one of Australia's biggest technology companies, reaching a market capitalization of $3 billion. His next company, Goggleboat, built the world's first all-plastic, seamless marine craft, again drawing on nature's spirals and curves—in this case, the sleek form of dolphins and whales. Light, strong, fuel-efficient, and nestable for low-cost shipping, his marine craft won Australia's national Design Award. Harman also designed and built a fifty-foot wooden sloop, which he sailed twenty-seven thousand miles throughout Asia.

In 1997, Harman founded PAX Scientific in San Rafael, California, to apply what he had learned about nature's efficiencies to seemingly mundane products: fans and mixers, propellers and turbines. He found a way to translate natural geometries into mathematical algorithms, which he then used to generate industrial products that are both incredibly beautiful and Spartan in their energy use. He was helped in the effort by Cascade Technologies, a consultancy set up by Stanford University turbulence experts. Harman's great innovation, according to Gianluca Iaccarino, an assistant professor of flow physics in the university's department of mechanical engineering, was his novel (and proprietary) strategy for "extracting the geometry" from nature's flows and forms. The PAX Scientific Web site illustrates just how deep the affinities are: a swirling cloudbank morphs into a fan; an elegant calla lily becomes a gleaming stainless mixer for use in water treatment plants.

One of the company's first commercial products is a little fan designed for refrigerator motors. Its blades at rest look as if they are already in motion, bulging and curving at the top like something out of Looney Tunes. Twenty-five percent more efficient than conventional fans, the PAX fan will cut overall energy use in refrigerators by 4 percent. That sounds small, until you consider that 15 million refrigerators are sold in the United States each year, which means that 4 percent could translate to 219,000 megawatt-hours of electricity not used. Fans are in every motor, compressor, and pump, and collectively use 15 percent of the total electricity consumed by U.S. industry; as PAX expands into these other areas of industrial design, the potential energy savings are enormous. Paul Hawken, cofounder of Smith & Hawken and coauthor with Amory Lovins of *Natural Capitalism*, is chairman and CEO of three companies that have licensed the PAX technology for production. Janine Benyus, author of *Biomimicry*, sits on PAX Scientific's board.

Serious Materials: Marc Porat, a legendary and controversial figure in Silicon Valley, wants a major piece of the $4.6 trillion global construction market. He already has a small one. Serious Materials, a company he founded in 2002, does a brisk business building soundproof envelopes made by its Quiet Solution division; an early customer was rap artist Snoop Dogg, who built a recording studio in his Hollywood home and sent his contractor to Porat's home to try to pay with a suitcase full of cash.

Porat's newest product, EcoRock, scheduled to be on the market in 2008, aims to displace conventional drywall, which is responsible for some 12 million tons of carbon dioxide emissions worldwide each year. Those emissions come primarily from the multiple "burns" required: first to dry gypsum (which comes wet from the mine), then to grind it, then to boil water to make a slurry, then to shape it and dry it again. The EcoRock drywall cooks itself through an exothermic chemical reaction. (An exothermic reaction generates heat energy, rather than requiring energy from an external source.) Because it needs no heaters, it produces no carbon dioxide. It is also stronger, cheaper, and lighter than conventional drywall, reducing the costs and emissions of transport.

In November 2007, Serious Materials closed a $50 million round of funding with Rustic Canyon and Foundation Capital, most of which the company will use to build the first EcoRock factory. Among the sites under consideration is one on the San Joaquin River in Stockton, California, right next to a factory owned by the nation's largest drywall company: U.S. Gypsum. Where that older factory has immense intake pipes for natural gas, EcoRock's factory will have none; its minimal power needs will come entirely from rooftop photovoltaics. "You have to be on the right side of the energy curve," says Porat. "If your costs rise with the price of energy, you're cooked. But if you can decouple, then you don't care what happens to gas prices."

Now Porat has an even more ambitious target: cement. Global cement production is responsible for about 5 percent of the world's greenhouse gases—more than all global air travel combined. Half those emissions come from calcium carbonate (limestone), which releases carbon dioxide when heated; half come from the 1,450°C (2,640°F) flame needed to "calcine" the cement, or drive off bound molecules of water so that it will be ready to react when new water is added. European manufacturers like Lafarge have invested millions of dollars trying to clean up the process; so far, they have cut emissions by only about 20 percent. The Portland Cement Association, which represents North America's manufacturers, says the U.S. industry will decrease emissions 10 percent below 1990 levels by 2020. Porat calls that target "insanity—stupid." He claims that his newest company, CalStar Cement, will use a reformulated recipe (replacing large amounts of calcium carbonate with fly ash, the fine residue left after coal combustion) and a proprietary heat-generating chemical reaction to eliminate *90 percent* of cement's carbon emissions. "The 10x change is what we're after, in all we do," says Porat. "Silicon Valley is a great place to lodge those dreams." Will they come true? The answer is unclear. But the Portland Cement Association acknowledges that cement can be made almost entirely of fly ash using "chemically unique, more rare fly ashes." And Porat has assembled an impressive team of technical advisers who are betting he may be right.

If Porat succeeds, he will have pulled off one of the great second acts in Silicon Valley history. His dazzling first act provided the centerpiece of "Diary of a Disaster," written by Richard Doherty and Apple cofounder Steve Wozniak. Porat's rapid rise began in 1976, when his Stanford doctoral thesis on the "information economy" was published; by 1980 he was director of the Aspen Institute, where he wrote, produced, and hosted a PBS documentary titled "The Information Society" featuring former Secretary of State Henry Kissinger and economist John Kenneth Galbraith. Porat's reputation as a visionary got him hired, in 1988, at Apple.

Long before Palm Pilot, instant messaging, Blackberry, and eBay, he conceptualized for Apple a handheld device that would access networks of information and an electronic marketplace and allow instant communication ("like the personal communicator badges they wore on *Star Trek*," Doherty recalls). Apple CEO John Sculley started to develop the device, then spun it off into an independent company with Porat at the helm. Porat brought along other Apple legends: in a March 15, 1998, article, the *San Jose Mercury News* likened the launch to "the Beatles reunion; the core Mac team minus Steve Jobs."

The utopian, new-age aura was thick at the new company, which Porat named General Magic on advice of the director of the psychic research program at Stanford Research Institute; he called employees magicians and exhorted them to "walk through walls." With tens of millions of dollars from Motorola, AT&T, and Apple, it was, recalls Doherty, "one of the highest flying start-ups in memory." The first device to use the General Magic software was the Sony Magic Link. And then the whole thing crashed, thanks to numerous fatal bugs in the device, a failure to reconcile the competing demands of General Magic's big corporate partners, and the launch of Netscape, which eclipsed it entirely. By 1995 employees and corporate partners were leaving; by 1996 Porat was gone.

Having largely vanished from public view for a decade, Porat seems immensely content to be back in the middle of Silicon Valley's next new thing, strolling in an elegant suit through his research facility in Newark, an industrial park north of San Jose. He takes every cell phone call (there seem to be no support staff and no land lines) and steps happily over the piles of sand and aggregate and around the cement mixers in the giant warehouse space next door. He still speaks with a prophet's eloquence about the global transformation that is now underway. Climate change—and the new energy revolution—are like tidal waves, he says. "When a tsunami moves through the ocean you can't see it. But when it reaches shore it becomes a hundred-foot wave."

Interface: In Ray Anderson's telling, his thirty-three-year journey through the land of industrial carpet manufacturing has been a kind of modern-day Pilgrim's Progress, taking him from the "City of Destruction," as John Bunyan had it, to "the world which is to come." For the first twenty-one years after he founded Interface in 1974—it has since grown to be the world's largest maker of modular carpet tiles—Anderson, who had played football for Georgia Tech, says he "never gave one thought to what we took from or did to the Earth." Then someone gave him Paul Hawken's 1993 book, *The Ecology of Commerce*, which argued for a new kind of enterprise that would become more profitable by reducing waste and energy use and its overall impact on the planet. The revelations Anderson found in those pages were like "a spear in the chest that has never left me. I realized the way I'd been running my company had been the way of the plunderer."

In the wake of that epiphany, Anderson made it his company's mission to "climb Mount Sustainability," fundamentally remaking its operations and choice of materials to achieve an ultimate goal of zero environmental impact. No fossil energy or petrochemical inputs, no solid waste, no air or water pollution. No footprint. Beyond that extraordinary goal, he committed to make Interface the world's first "restorative company," contributing more than it was taking from the natural world, publicly documenting its progress so others could follow the example. Anderson's allegorical way of speaking can seem overwrought on the page, but it is impossible to ignore how much he has already achieved. Since 1995 his company has reduced greenhouse gas emissions by 56 percent, saving more than $300 million while increasing sales 49 percent.

Anderson's strategy has been to "burn bridges," abandoning destructive and wasteful practices. One such abandoned process is the printing of patterns on carpet tiles. Dye printing requires huge amounts of energy and water: for applying the dyes, steaming and fixing the colors, washing out the excess, drying the carpet, and treating the excess water. Anderson's team figured out a way to use the tufting machine—a kind of huge sewing machine with hundreds of needles that insert the loops of yarn into jute, cotton, or synthetic backing—to create the patterns, decreasing water and energy use by 90 percent.

Another innovation came in the carpet pattern itself. The company's top-selling pattern, called Entropy, mimics the disorder of a forest floor with its strewn leaves, pebbles, and twigs. That randomness means that the pattern needn't match up from tile to tile, but can be laid in any direction, eliminating the huge amounts of scrap normally generated at installation. It means few tiles are rejected at the factory: imperfections get lost in the wandering variations of color. It also means the carpet lasts a long time, because worn or stained tiles can be swapped out without replacing the rest.

The company's clear mission, says Anderson, has had a galvanizing effect on employees. He tells a story about a visitor to one Interface factory who approached a lift truck operator and asked, "What do you do?" The young man stopped his machine and climbed down to answer: "Ma'am, every day I come to work to save the earth."

BSST: The C2 Climate Control doesn't look like a revolution. Just ten inches tall and made of white plastic, the little space heater—designed to heat the area immediately around the user's body—looks more like a flat-bottomed blow dryer, or a retro telephone. In fact, the C2 is one of the first commercial products to use advanced thermoelectrics: solid-state devices that turn heat to electricity, or electricity to heat. Produced by Herman Miller with cutting-edge technology from a California company called BSST, the C2 uses 90 percent less energy than a traditional space heater, without heating coils, chemicals, emissions, or fire danger.

Like so many of these advanced technologies, thermoelectrics has its roots in the nineteenth century. In 1821 a German-Estonian physicist named Thomas Johann Seebeck discovered that certain materials, like lead sulfide, produced an electrical current when one side was hotter than the other. That is, heat became electricity without first being used to make steam and then drive moving parts. The reverse was also true: if electricity was applied to these same materials, one side became hot and the other cool.

For a long time, that conversion of heat to electricity or vice versa was deemed too inefficient, and the materials too expensive, to be commercially interesting. But recent advances by BSST have changed that. The company is now working with the Departments of Energy and Defense, materials scientists at national laboratories, and private companies to expand applications for this solid-state heating, cooling, or power-generating device.

The first commercial thermoelectric product brought to market by BSST's parent company, Amerigon, was a car seat that warms or cools its occupant, now available in more than twenty models of cars. The C2 was next, coming to market in January 2008. Now BSST is working with BMW North America and other partners to develop a product to convert waste heat from vehicle exhaust into electrical power. In current automobile engines, about 70 percent of the energy value in the gas is lost to the atmosphere as waste heat; capturing that heat and converting it to electricity to run the car's air-conditioner, heating, and electrical systems, according to BSST cofounder Lon Bell, could improve fuel economy by 10 percent. BSST is also working with Carrier, the UTC subsidiary involved in the Chena Hot Springs geothermal power project profiled in

Chapter 7, to develop solid-state refrigerators and air-conditioners. Ultimately they envision industrial-scale thermoelectric heating and cooling systems, and large-scale systems to make electricity from the waste heat that is produced in abundance in every building and factory and power plant in America. When that happens, according to the Oak Ridge National Laboratory, "the impact on America's fuel consumption, and the resulting impacts on security and the environment, will be of enormous and lasting value."

IBM: Every year, IBM spends more than $6 billion on research and development, ranking it among the top-ten corporate R&D investors in the world. An increasingly big chunk of that money is going to reducing energy use and carbon emissions, both its clients' and its own. In 2007, "Big Blue" launched a new business unit, called Big Green Innovations, tasked with finding new ways of applying the company's deep expertise in computational modeling and visualization, and the immense power of its supercomputers, to all kinds of environmental problems—including helping enterprises like Wal-Mart make their way through the vast thickets of data on emissions throughout their supply chain. In May 2007, IBM debuted "Project Big Green," a billion-dollar-a-year initiative to redesign data centers to cut energy use up to 40 percent. It was its own first customer: according to Wayne Balta, IBM's vice president of corporate environmental affairs, the company will double its computing capacity in the next three years "without using another electron to do it." Pacific Gas and Electric was another early customer: the utility switched from the three hundred servers it had been using to six IBM mainframe-based "virtual servers"—single computers programmed to do the work of dozens—reducing energy use in its data centers by 80 percent.

In October 2007, IBM announced that it had developed technology that permits recycling of silicon wafers, helping ease the silicon shortage that has plagued developers of solar power. And yet another new service builds on the expertise IBM has accumulated designing weather computers for the National Oceanic and Atmospheric Administration. Called Deep Thunder, it provides real-time, high-resolution weather predictions (down to a square mile) to help companies stay efficient no matter what the sky brings down. "Suppose you were in the business of moving cargo or you were a forest products company sending heavy equipment into the woods," says Balta. "Wouldn't you want to be able to anticipate the weather? Solving climate change will require new technologies. But the world just isn't going to stop working how it does." The company is also developing water management technologies, including a smart irrigation system that combines soil sensors with Deep Thunder's weather forecasts to ensure minimal water use: it will save not only up to half the water now used in irrigation, according to Balta, but also the enormous amounts of energy used to move that water around. That could mean big savings: fully 90 percent of all electricity used on farms goes to pump water for irrigation; the California Energy Commission estimates that almost 20 percent of the state's total electricity demand is associated with water use.

WHAT CAN BE DONE to accelerate adoption of the new wave of energy-efficient technologies? A cap-and-trade system will help, since it will increase still further the value of energy savings. But in trying to affect the decisions of millions of individual consumers, a thicket of other obstacles may need to be pruned away, through policies that complement cap and trade. Until now, this book has focused on the overarching market failure, what economists call the "tragedy of the commons"—the degradation of publicly owned resources in the absence of private incentives to protect them.★ But there are more localized market failures that need to be addressed as well.

One has to do with information. Consumers often lack good data about the differences in energy efficiency among products, and the resulting energy savings. When choosing between a more expensive energy-efficient air-conditioner and a cheaper energy guzzler, most will focus on the upfront savings—unless they are clearly and persuasively informed about the money they will get back in lower electricity bills over the unit's lifetime.

Another market failure concerns what economists call "principal-agent problems," which arise when the person who decides what level of energy efficiency to install is not the person who saves the money. Think of the real estate developer who chooses inadequate insulation because homeowners rarely make that the basis of their purchases, even though better insulation would save more money than the extra cost of installing it. Or the landlord who lacks incentive to fix drafty windows or install more efficient appliances because the tenant pays the utility bills. Or the television manufacturer who builds a product that hogs power even when it is in "sleep" mode. In these cases, simple government approaches can usefully buttress a cap-and-trade system, including improving building code standards,

---

★ The metaphor comes from a 1968 article of the same name in the journal *Science*, in which Garrett Hardin evoked the image of herdsmen tending livestock on a common pasture. Collectively, they would benefit from avoiding overuse of the commons, protecting its future productivity. But each individual's self-interest leads him to add more animals to the pasture, since the owner alone reaps the benefits of a larger flock, while he shares the costs of overgrazing with the other herdsmen. Thus the "tragedy": individual rationality leads to an outcome that leaves everyone worse off. In the case of climate change, the atmosphere is the public commons being destroyed by a lack of private incentives to protect it. A cap-and-trade mechanism provides those private incentives and remedies the tragedy.

energy labeling, and efficiency standards for appliances. These are narrow-gauge approaches, not a comprehensive solution. If carbon trading provides the overarching framework—the economic signal that drives innovation and reorients market decisions—standards can ensure that the signal is transmitted to the final decisionmaker.

THE THIRD GREAT NEAR-TERM OPPORTUNITY to cut greenhouse gases is, of course, automobiles, which add another twenty pounds of carbon dioxide to the atmosphere with every gallon of gas they burn.

One camp is betting that electricity will be the answer. A 2006 report from AllianceBernstein, a global investment firm with almost $800 billion in assets under management, deemed it "inevitable" that cars will soon run primarily on electricity, sending internal combustion engines the way of the coal-fired steam train. The main value of electricity is that it is a carbon-free energy carrier—the only one known, apart from hydrogen. Even accounting for the largely carbon-intensive generation of electricity in the United States, battery-powered cars are responsible for only about half as much carbon dioxide per mile as comparable gasoline cars, and the picture is even better in regions with a relatively clean power mix, such as California. So while electric cars will not make a major dent in global warming pollution until battery costs come down and U.S. electric power generation relies far more heavily on solar, geothermal, wave, and wind, they will eventually become very important indeed.

The history of electric cars is as old as man's use of electricity, but the latest renaissance of the technology—which has a fervently passionate following—has taken several new turns. The most buzzed about has been the entry of Silicon Valley into the game, primarily in the form of the sleek Tesla Roadster, the gorgeous, spectacularly powerful electric sports car that has already been pre-purchased by everyone from actor George Clooney and Governor Arnold Schwarzenegger to Google cofounder Sergey Brin, despite its $100,000 price and the fact that its delivery date has been repeatedly delayed. Funded by the wildly visionary Elon Musk (who sold PayPal to eBay for $1.5 billion, has a private rocket-building company called SpaceX, and signed a $278 million contract with NASA to design and build the successor to the Space Shuttle), Tesla Motors is attempting to remake a car company in the image of an advanced-technology company. It powers the car with almost seven thousand lithium-ion batteries lifted straight from the laptop computer business (the batteries, the company says, are recyclable—and the cost of recycling is built into the purchase price). It conducts "open-source" research and development: on his online blog, founder Martin Eberhard asked electric vehicle enthusiasts to download Excel spreadsheets and record information about their homes' circuitry and appliance loads. It aims at the top end of the market, believing that—as with cell phones and plasma TVs—breakthrough technologies in cars will start out expensive and trickle down. And it is convinced that it can topple the industrial-age dinosaurs. In July 2007, four months after SpaceX launched the Falcon 1 rocket (on which Musk had been chief designer despite his total lack of formal training), the thirty-six-year-old Tesla backer told *Business Week*, "Silicon Valley is the best in the world at everything it does."*

At first glance, the Think City car from Norway seems a more familiar kind of electric car: small (two seater), short range (up to a hundred miles), and relatively low performance (sixty miles per hour tops). But Think—owned for a time by Ford, which invested $100 million in its safety features and design—is also a creature of

---

* *Business Week*, July 30, 2007.

the information age. Its charmingly rumpled CEO, Jan-Olaf Willums, has been brainstorming, along with Larry Brilliant, head of Google.org, and others on the Google team, on the car company of the twenty-first century. They decided, in the end, that it is not a car company at all, but a service company offering "mobility"; it may organize car sharing among its customers, for instance, or lease batteries rather than sell them, allowing people to upgrade their cars in much the same way they upgrade computer operating systems. Like Dell, Think will likely sell cars online, built to order; among the optional equipment choices may be a home rooftop solar system, so customers can buy a car and its zero carbon energy at once. It will equip each vehicle with WiFi, GPS, and Web-based maps and traffic information. Information technology, in other words, will be used to provide brains rather than brawn (an advance that will be just as valuable in gasoline-powered cars). A "networked car," communicating with other vehicles or the city itself, could be smart enough to find its way around traffic jams. It could consult with curbs or parking lots that know when they're full, and get directions to a more promising spot, including one with plug-in power.

Over the horizon, more promises await. The plug-in hybrid Chevy Volt from General Motors, for example, may hit the market by 2010. The Volt's electric motor will take it up to forty miles on a single six-hour charge, covering most Americans' daily commute. For longer trips, a small internal combustion engine will kick in, though only to recharge the batteries. With the help of that engine—able to burn gasoline, biodiesel, or ethanol—the car will go up to six hundred miles and get the equivalent, GM says, of fifty miles per gallon. The Volt is still in development, its battery remaining the biggest sticking point. But, says Robert Lutz, GM's vice chairman for global product development, "We would not be devoting the considerable investment in engineering if we weren't confident that it could be done." In April 2007, GM unveiled a variation on the hybrid theme at China's Shanghai Auto Show: Chevy Volt Hydrogen, which replaces the backup internal combustion engine with GM's fifth-generation hydrogen fuel cell.★

California-based car company ZAP, which has been making small, low-speed city cars for more than a decade and recently signed manufacturing agreements in China, says that its lightweight aluminum ZAP-X, with high-tech hub motors in every wheel, will deliver 644 horsepower, reach speeds up to 155 miles per hour, and go 350 miles on a ten-minute recharge. Lotus Engineering, the British sports car manufacturer, designed the car, whose lightweight body is combined with "exceptionally strong and stiff structural rigidity." (*Automotive News* notes that earlier questions about aluminum's safety have been put to rest by the proliferation of high-performance, lightweight vehicles in auto racing: "Deaths have decreased significantly while the weights of the vehicles have gone down progressively. . . . In this case, lighter is markedly safer.")† By eliminating the engine, transmission, drive train, and conventional brakes, ZAP says, it will open up space beneath the floor for a giant lithium-ion battery pack. A still newer company, PML Flightlink, recently got $40 million to develop its own hub motors to be incorporated into car wheels, which it says will recapture 80 percent of the braking energy, using it to recharge the batter-

---

★ The most common type of fuel cell uses a catalyst to separate a fuel (typically hydrogen) into positively charged protons and negatively charged electrons. The protons can travel directly through the membrane that separates the two electrodes. But because the membrane is insulated, the electrons have to go the long way around, through an external circuit, creating an electrical current. Fuel cells have so far remained too expensive—and the problems with hydrogen storage too knotty—for the technology to go mainstream for portable use.

† *Automotive News*, October 17, 2005.

ies. (The idea of putting an electric motor in each hub was first invented in 1897 by Ferdinand Porsche.) And a few innovators are working on cars that run on compressed air: batteries run the compressors; when released, the air drives the pistons.

The Achilles heel of electric cars—the main technological impediment to large-scale production—has long been the battery. Now, even batteries may be poised for a quantum leap: in storage capacity, quick recharge, lifespan, and affordability. "Batteries have been just around the corner for thirty years," says Steve Pacala, director of the Princeton Environmental Institute, and a member of the Environmental Defense Fund board of trustees. "This time they really are just around the corner."

Most innovators are modifying the chemistry of lithium-ion batteries to make them cheaper (the battery pack in each Tesla Roadster costs $20,000), smaller, lighter, and safer (in computers, lithium-ion batteries have been known to go up in flames). A123Systems, an MIT spin-off, and Altair Nanotechnologies of Reno, Nevada, are using nanotechnology to increase battery stability and performance and reduce cost, and both have deals with car companies. California's Air Resources Board calculates that with mass production, the price of lithium-ion battery packs will fall below $4,000, cheap enough for a mass-market car. A123 has more than $100 million in funding and 250 employees in the United States and Asia, is already supplying batteries for power tools to Black & Decker and DeWalt, and is one of two companies developing batteries for the Chevy Volt.

More radical new storage approaches are also being explored. A company in Austin called EEStor has stirred intense curiosity with its new ultra-capacitor technology for cars, not least because it has maintained a silence worthy of Greta Garbo. When in 2007 the company was awarded its first key patent, the Associated Press reported that "energy insiders spotted six words in the filing that sounded like a death knell for the internal combustion engine."★ The six killer words: "technologies for replacement of electro-chemical batteries." Because a conventional battery relies on chem-ical reactions to store energy, it can be slow to charge and discharge. An ultra-capacitor instead sandwiches a chemical compound between thousands of wafer-thin metal sheets: charged particles stick to the metal sheets and move quickly across the material. In EEStor's case, the proprietary sandwich is made from barium tita-nate. The claims made for the device are extraordinary: that it will charge in five minutes, power a car for five hundred miles, weigh less than a hundred pounds, tolerate extremes of cold and heat, cycle through millions of discharges and recharges without giv-ing out—and be powering Canadian-manufactured ZENN Motor cars by 2008. Joseph Perry of Georgia Tech says his researchers have used a similar approach to double the amount of energy a capacita-tor can hold, but that EEStor is claiming a 400-fold improvement; Perry is skeptical, but hopes to be proved wrong.

If the problems with batteries can be solved and electric cars or plug-in hybrids finally manage to go mainstream, that will pre-sent another game-changing possibility: the use of car batteries as a huge distributed storage system for off-peak and renewable energy. A consumer would "fill up" at home, using a standard wall socket. She would no longer need gas for local errands, and the cost of driving each mile would drop below 4 cents, equivalent to one-buck-a-gallon gas. On top of that, she would get paid by the local utility for agreeing to plug the car in during certain hours of the day, thus providing standby power to the electric grid in case it should need a quick burst of electricity; Hal LaFlash, director of renewable-energy policy for PG&E, predicts payments from this vehicle-to-grid (V2G) system might reach several thousand dollars

★ Associated Press, September 4, 2007.

a year. The car battery would also serve as backup power should the grid go down, storing enough electricity to run a whole house (minus the air-conditioner) for two days.

Pacific Northwest National Laboratory calculates that there is enough excess nighttime generating capacity nationwide to charge 84 percent of the 198 million cars, pickups, and SUVs on the road today. Ideally, the energy charging those cars would come from carbon-free sources, which would reduce almost to zero the greenhouse gas emissions per mile. But a joint study by the Electric Power Research Institute and the Natural Resources Defense Council found that even if a plug-in vehicle got all its electricity from coal-fired plants (the U.S. electricity grid is about 50 percent coal), it would still emit only two-thirds of the greenhouse gases released by a conventional car. Over the next thirty years, the study concluded, widespread adoption of V2G could eliminate 450 million tons of carbon dioxide annually, the equivalent of retiring a third of the current fleet.[*]

Several U.S. utilities have begun developing the infrastructure necessary to support V2G, with PG&E in the lead. LaFlash says that rather than build new plants to meet marginal peak demand or pay gas-fired generators to quickly turn off and on, he would much rather use electric cars to shave those peaks. One thousand vehicles would hold a megawatt of power; four hundred thousand vehicles would hold the equivalent of a small coal-fired plant. Princeton's Steve Pacala explains that a utility's generating options become far greater if it does not have to be 100 percent reliable every minute of every day (that is, if it does not need enough extra power turned on at all times so that no matter who flicks a switch, the light comes on), but instead can fill in gaps from a distributed storage system. And while it takes $1,500 to build a kilowatt of capacity at a

---

[*] "Environmental Assessment of Plug-in Hybrid Vehicles," July 2007.

power plant, he says, it costs just $15 to make a kilowatt of capacity in a car. PG&E has also announced plans to buy used electric-car batteries once they have outlived their usefulness for transport and install them in basements of offices and substations to store more green energy.

PG&E is now working with Tesla on ways to remotely control charge (and discharge) times. With Google, the utility has installed a test recharging station to demonstrate the bidirectional flow of electricity between the car batteries (made by A123Systems) and the grid (though at the Googleplex, most of the power is coming from the solar panels installed by Energy Innovations); PG&E sends wireless signals to the cars to determine their state of charge. More than one observer has noted the potential business opportunity for Google in a V2G system: providing the data and network management infrastructure to figure out where millions of cars are plugged in, matching the power level in their batteries with the grid, developing meters that can track the energy going in and out of the cars, and figuring out what to charge or credit the car owner. With such a system, says one prominent venture capitalist, "PG&E will cut ExxonMobil at the knees."

NOT EVERYONE BUYS THE IDEA that California start-ups are going to topple Detroit. Though the idea of a visionary remains perpetually fascinating, these new car companies may never achieve the reliability and affordability demanded in a mass-market vehicle. Tesla Motors has raised $105 million, but the auto industry typically spends $500 million to $1 billion—plus all its other corporate resources (wind tunnels, modelers, parts suppliers)—to develop a new car, even one using conventional technology. As Detroit likes to remind everyone, the last successful car start-up was Walter Chrysler's namesake company in 1925.

The challenges Tesla faces became clearer in October 2007,

when the company acknowledged that it had fallen a year behind schedule, undone by the complex logistics of getting components produced in Thailand, Taiwan, and the United States to arrive at the right time at the assembly plant in England. They replaced Martin Eberhard as chief executive, and in December ousted him entirely. "Silicon Valley engineers find it easy to think they know everything and Rust Belt companies don't know anything," Eberhard had told the *Wall Street Journal*. "More often than not the knee jerk reaction, that these guys in Detroit don't know what they are doing, is wrong."★

Though there is no question that we need the broadest possible research and development portfolio for the car of the future, some observers worry that the dazzling promises made for electricity (and for hydrogen, which for a time claimed the spotlight and still has its acolytes) have effectively distracted policymakers from the carbon-cutting opportunities immediately at hand. That carmakers could dramatically cut carbon emissions without *any* technological breakthroughs has been made patently clear by industry leaders elsewhere in the world. From 1990 to 2005, Toyota cut average fleetwide emissions not primarily by introducing the Prius, but by incrementally improving efficiency in the Corolla. BMW cut its fleet average 12 percent with efficiency improvements and the introduction of the MINI Cooper.

Even the leading car companies could go further still, says Axel Friedrich, head of the transport department at Germany's Federal Environmental Agency. In 2007 he hired university engineers to rebuild a Volkswagen Golf with off-the-shelf parts: they cut emissions 25 percent while keeping horsepower intact. A study by the National Academy of Sciences estimates that 2 billion gallons of fuel

---

★ *Wall Street Journal*, July 30, 2007.

a year could be saved in the United States just by reducing the rolling resistance of tires 10 percent, with no compromise on safety.

Without any major breakthroughs, vehicles that are little different from today's could use one-third the energy per mile, says John DeCicco, Environmental Defense Fund's specialist in automotive strategies. That alone would radically reduce greenhouse gas emissions. If those cars ran on a biofuel made from renewable feedstocks with one-fourth the lifecycle greenhouse gas emissions of today's gasoline, then the emissions per mile would be one-twelfth what they are today. It was the feasibility of such options that, in September 2007, caused Vermont U.S. District Judge William K. Sessions to reject manufacturers' challenges and rule that they could meet California's new standards, requiring carbon dioxide emissions in new cars to be cut about 22 percent in the first phase (2009 through 2012) and 30 percent in the mid-term phase (2013 to 2016).★ Given the expected doubling by midcentury of vehicle miles traveled in the United States, however, the country will have to go much further—reducing automobile emissions about 80 percent.

One of the most promising steps in that direction has emerged from the Sloan Automotive Laboratory at MIT. After several decades as head of the university's fusion center, Professor Daniel Cohn realized he wanted to work on something that would come to fruition in his own lifetime. With his old friend John Heywood, who runs the Sloan lab (he is often called "the Yoda of cars"), and Leslie Bromberg, one of the world's great computer modelers, he figured out a way to combine existing technologies to make a small internal combustion engine perform like a big one, improving fuel efficiency by 25 percent. The key is overcoming knock, the spontaneous explosions that occur in overheated cylinders and limit the

★ In December 2007, the EPA denied the waiver California needs to enforce its standards. The state has challenged the ruling.

amount of power carmakers can get from each piston stroke. Cohn and his colleagues discovered they could cool the overheated cylinders by spraying a bit of ethanol directly into the combustion chamber before the spark; ethanol takes a lot of heat to evaporate, so when it does it cools the cylinder far more effectively than water might, and without degrading the fuel mix. They called the technology—and the company they founded—Ethanol Boosting Systems (EBS). A small, light, four-cylinder engine equipped with an EBS, they claim, could power a giant truck. The necessary modifications—a second fuel tank for ethanol, an injection system, and a small exhaust-driven turbocharger (to compress more air into the cylinders)—will cost about $1,500, far less than the extra price consumers now pay for a hybrid (about $4,000 in 2007). With increased engine performance, the technology saves three gallons of gas for every gallon of ethanol consumed, reducing net carbon dioxide emissions 20 percent.

Even without using ethanol boosting or other exotic approaches, gasoline engines could perform far more efficiently than they do today. In a January 2007 paper for the Society of Automotive Engineers, for example, Heywood and an MIT graduate student described how advanced turbocharged gas engines could use 40 percent less fuel than conventional engines. Heywood has argued vigorously for changes in government policies—including stricter Corporate Average Fuel Economy (CAFE) standards—to help "push and pull this technology into the marketplace and ensure it is used." The United States has lagged other developed countries on this score. While the whole European fleet averaged 44.2 miles a gallon in 2007, and Japanese cars edged above 45 miles per gallon, the U.S. Congress finally agreed to require American cars and light trucks to average 35 miles a gallon—by 2020.

Perhaps more than any other area, the transportation sector

requires integrating multiple policy approaches. Sector experts talk about the need for policies to motivate all of the actors—fuel producers, automakers, shipping firms, land-use planners, infrastructure providers, and consumers—whose choices and actions together determine transportation greenhouse gas emissions.

Like energy-efficiency standards for buildings and appliances, CAFE standards can play a role in lowering carbon emissions. But the bigger transformation will depend on enacting the overall framework of a carbon cap-and-trading system. The most straightforward way to bring transportation into a U.S. carbon market would be to enact an "upstream" cap—that is, to cap the carbon from fossil fuels at the refinery gate or port of entry. Each ton of carbon in the petroleum that goes into gasoline and diesel fuel translates into 3.67 tons of carbon dioxide coming out of the tailpipe—and can therefore be traded on an apples-to-apples basis with tons of carbon dioxide from a power plant or factory smokestack.

Such an upstream cap would cover all fossil fuels—not only petroleum but also liquid fuels made from coal and natural gas. It would not, however, apply to biofuels like those Amyris, Verenium, and GreenFuel Technologies hope to produce, which would therefore gain an important competitive advantage. (Of course, the biofuels manufacturers will have to pay the carbon premium on any fossil fuels they use to grow and convert their crops, so their net carbon emissions will also be reflected in their price.) As the true costs of the global warming impacts of gasoline show up in the price at the pump (carbon at $25 a ton in the global market would translate to about 25 cents more per gallon of gas), consumers will have a new reason to press for cars that are both more efficient and able to use a variety of fuels—a pressure that will be felt by U.S. automakers.

————

IT IS NOT JUST THE GUTS and the brains of the car that
can be remade to reduce energy use and emissions, but also the
body. Bill Gross of Idealab in Pasadena has a car company, Aptera
Motors, that is focused on radically streamlining aerodynamics and
reducing weight to get a car that can go 250 miles on a gallon of
gas. "Think of the tremendous force on your hand when you stick
it out the window of a fast moving car," he explains. "It's the same
with the car itself—you use five horsepower to move the car, and
ninety-five horsepower to move the wind out of the way." The two-
passenger prototype is beautiful—as smooth and slippery looking
as a dolphin, but big enough, according to Aptera's Web site, "for a
couple of seven-foot surf boards and associated beach accessories."
The detailing is extraordinary: for instance, extended mirrors eat
up ten miles per gallon, so instead they have set cameras flush into
the body that transmit to LCD screens. The Web site notes that
"based on our wheel layout and our weight," the vehicle is actually
categorized as a motorcycle—but a motorcycle engineered with
safety firmly in mind: "We chose not just to meet many of the
specs for passenger vehicles . . . we chose to exceed them whenever
possible." Gross thinks they will sell about ten thousand cars a year,
but his real goal is to affect mainstream car design. Aluminum giant
Alcoa is also engaged in efforts to radically reduce the weight of
vehicles. In 2007 it formed a partnership with Zhengzhou Yutong
Bus Company to build aluminum buses for the 2008 Olympics,
which will cut overall weight by as much as 20 percent.

The biggest opportunities for emissions reductions may be in
the developing world, where demand for vehicles is growing fastest. A hundred million motorcycles, scooters, and three-wheel
"tuk-tuks" in Southeast Asia are powered by two-stroke engines,

each of which produces as much pollution as fifty cars. Envirofit International, a nonprofit organization in Colorado, has developed a fuel injection retrofit kit that cuts fuel use and carbon dioxide emissions from those two-stroke engines by 35 percent, with still greater reductions (up to 90 percent) of the conventional pollutants that cause cardiac and respiratory illness. By 2010 the organization will have retrofitted three thousand motorcycle taxi engines in the Philippines—eliminating 3,000 tons of carbon dioxide a year and saving drivers $1.4 million in fuel costs. Envirofit is now collaborating with microfinance groups to enable drivers to buy the $350 kits with loans, paying them off in the first year with the $500 they save on fuel.

THE LAST ARENA of automobile-related innovation is focused on reducing the amount of driving—currently projected to increase in the United States by 60 percent over the next two decades. These increases threaten to overwhelm gains won through more efficient vehicles alone.

Smarter real estate development is one important solution. A 2007 study by the Urban Land Institute found that shifting two-thirds of new growth to compact patterns would save 85 million tons of carbon dioxide annually by 2030. California's San Joaquin Valley recently put in place financial incentives to encourage developers to design subdivisions that reduce the need to drive. Going forward, policymakers may have to require more accountability for the carbon dioxide impacts of transportation and infrastructure decisions. Simply using the most advanced building technologies can make a big difference; "cool paint," for instance, which reflects infrared light and therefore cuts heat absorption by half, can dramatically reduce air-conditioning loads.

Getting incentives right for drivers and commuters is also critical. London has cut carbon dioxide emissions from vehicles in its city center by some 20 percent with congestion pricing—fees charged for entering the center during peak hours. On a growing number of roads from Singapore to San Diego, tolls are being adjusted by time of day to ensure that roads operate efficiently without congestion, with revenues funding better transit services. In Yorkshire, England, a private road developer's earnings are tied to how well traffic moves. In Germany, trucks with dirty engines pay higher tolls.

Pay-as-you-drive insurance, which rewards drivers who reduce their annual mileage with lower premiums, could reduce Americans' driving by as much as 15 percent. Information feedback systems are already optimizing passenger and freight routing: IBM is providing integrated, real-time information on trains, buses, highways, and airlines to allow travelers and shippers to avoid bottlenecks and breakdowns. Used widely around the globe, such innovations could save 600 million tons of carbon by 2030.

A move toward greater efficiency in cars, combined with the advent of biofuels and better planning, says Pacala, will profoundly alter the economics of oil, the single most important determinant of our energy security. "We've had a nonmarket cartel controlling petroleum for decades," he explains. "With that one bloc controlling 45 percent of the total and all the spare capacity, they have had a complete monopoly on price and trapped us in a rigged game for thirty years."

A global cap-and-trade system would build a substantial industry providing alternatives to oil, and thus would cause the price of oil to fall to the price of alternatives—about $40 to $50 a barrel. "The extra dollars per barrel they're getting, the transfer of rents resulting from that monopoly market, adds up to about $1 trillion a year, which is more than enough to pay for climate solutions," says Pacala. "All it would take to get petroleum down to that free-market price is opening up a gap, not a very big gap—just a couple of million barrels a day—between the supply and the demand. That would break their capacity to sustain those price controls and generate enormous savings in consuming countries, which could be used to finance global warming solutions."

CHAPTER 10

A World of Possibility

Beyond even the most dramatic transformations discussed so far lie technologies that will require huge leaps of science (and probably of faith), but could change the game like nothing ever seen before. Some have not gotten much beyond a lab experiment; those that have remain closely guarded. Still, they map a sense of the extraordinary flights of imagination and invention possible, if we truly commit ourselves to confronting climate change.

Nuclear Fusion: Tri Alpha Energy in Foothill Ranch, California, is out to accomplish what every physics graduate student knows is impossible: building a hot nuclear fusion reactor that not only avoids the destructive flow of neutrons but also (unlike every other fusion experiment so far) might actually work at a practical scale. If it does, it will transform the world "on the scale of aliens beaming down to the planet," says Charles Byrd, a physicist who oversees technology investments at MissionPoint Capital and who investigated the technology several years ago for GE.

Fusion is what powers the stars. Unlike the fission used in current nuclear plants—which splits atoms (and produces long-lived radioactive wastes)—fusion pushes atoms together, compressing two or more light atomic nuclei to form a single heavier one. The sun, for instance, continually presses four hydrogen atoms together to create a single helium atom. That helium atom weighs less than the four original hydrogen atoms. In accordance with Einstein's most famous formula ($E=MC^2$), the change in mass is transformed into a burst of energy.

The hard part, here on earth, is first to heat the atoms to the million-degree temperatures needed to force them together, overcoming the natural repulsion between positively charged nuclei, and then to keep that super-hot gas confined. The big research centers—including ITER in France (a collaborative effort among the European Union, the United States, Russia, China, India, Korea, and Japan), and the U.S. national labs (including Lawrence Livermore and the Princeton Plasma Physics Laboratory)—have spent many decades and tens of billions of dollars trying to solve those problems. Yet to date, humans have managed to harness fusion only for the hydrogen bomb (which needed fission to get it going).

The Tri Alpha facility, led by eighty-two-year-old Norman Rostoker, a professor emeritus at University of California, Irvine, who has studied fusion since the 1950s, is a universe away from those multibillion-dollar labs, though many of the scientists involved are refugees from those tonier settings. When Byrd led a group of senior GE executives—who were considering investing in the company—to meet the scientists, they were led into a conference room furnished with folding plastic tables and collapsible lawn chairs. GE was sufficiently impressed that it did some research and development for the group, developing several magnets; while GE was still deciding whether to invest, Microsoft founder Paul Allen stepped up with millions of dollars. (Allen has consistently bet on the small team against the huge, government-funded enterprise: he also funded the *SpaceShipOne* team, which in 2004 designed, built, tested, and launched a manned space mission for just $25 million.)

Instead of the radioactive tritium that is fused with deuterium in most fusion reactors, Tri Alpha aims to fuse protons with a small block of nonradioactive boron-11. (Boron-11 is an isotope of boron, an abundant element used in the production of fiberglass, among other things.) Called p–B11 fusion, the process requires overcoming an energy barrier five times higher than the barrier between a deuterium and a tritium nucleus, a technical challenge that has so far defeated the world's most well-funded centers of fusion research. But it also promises to be vastly safer than alternative technologies. Rostoker described it in the November 2005 issue of *Florida Physics News*, an alumni newsletter from the University of Florida: The fast beam of protons chases the slow beam of boron-11, "rear-ending" it with the energy at which the rate of fusion is highest. Instead of neutrons, which can damage reactor walls and create radioactivity, the process produces charged particles; those particles are guided into a "Direct Energy Converter," where their kinetic energy is turned directly into electricity. The end product is helium, with no residual radiation. And while the billion-dollar reactors cover square miles, this reactor—if Tri Alpha ever succeeds—will be small and modular, like a gas-powered turbine. A block of boron-11 the size of a case of wine could power a 100-megawatt facility. It will still require staggeringly high temperatures, which in turn will necessitate containment. Rostoker describes their containment vessel as a cylinder that rotates on its own axis inside a magnet, producing "a magnetic field that closes in upon itself: a kind of self-confinement" and has "no danger of runaway reactions or explosions."

Scientists have discussed the possibility of such "aneutronic fusion" since the 1950s—and consistently dismissed it as impossible. When the Tri Alpha team published its initial findings in *Science* and *Nature* a decade ago, a flood of mail from nuclear and plasma physicists mocked them as "complete morons," in Byrd's telling. They have not yet gone up in smoke, however. In 2007 they raised $40 million in additional venture funding from Goldman Sachs, Venrock, Vulcan Capital, ENEL Produzione, and PIZ Signal. Though they have no Web site (and decline most interviews, including one for this book), CEO Dale Prouty did break the silence long enough in June 2007 to tell *Red Herring* that he expects it will take his company "not 15 to 20 years, but not 3 to 5 either" to go from the research stage to power generation.

Solar Fuels and Viral Batteries: Nate Lewis at Caltech thinks neither batteries nor biofuels will provide a low-carbon, mass-market solution to the problem of mobile power for cars and trucks and planes. So he is attempting to invent a kind of artificial photosynthesis, turning solar energy directly into liquid fuels. That would solve the solar storage problem as nature does—by breaking and making chemical bonds—but at much higher efficiencies than plants are able to achieve.

With funding from BP, Lewis is developing a membrane, or "artificial leaf," that like a real leaf will do three things: capture sunlight, convert it to a wireless current (as Daniel Nocera, an MIT chemist involved in similar experiments, puts it, "leaves are buzzing with electricity"), then store that energy in chemical bonds. Because they will not have to do anything else (grow, fight off pests, reproduce), Lewis says, his leaves can be optimized as energy conversion and storage machines.

For their first job, capturing light, Lewis will replace the chlorophyll used by real plants with intensely colored metal oxides similar to the pigments in paint. For their second job, converting light to electricity, he will add nanoscale semiconductors. As in other solar cells, these photovoltaic semiconductors will turn photons into excited electrons and holes. For the final step, converting that current to chemical energy, he will supply water and catalysts. As in real photosynthesis, the holes and electrons will split the water into oxygen and hydrogen and attach the hydrogen to carbon dioxide to make fuel. The fuel made by photosynthesis is sugar; Lewis is experimenting with making methane, methanol, or hydrogen.

His results in the lab have been promising. In 2006 his artificial leaves were making hydrogen and oxygen from water about five times more efficiently than plants do. By 2007 he briefly hit conversion efficiencies over 20 percent. The problem, he says, is the stability of the system: colors strong enough to absorb large amounts of light have deteriorated quickly.

Angela Belcher, an MIT professor of biological engineering and materials science, has not given up on batteries, but has invented a radical new way of making them. She has genetically engineered a virus, the M-13 bacteriophage, to grab conductive metals—cobalt oxide and gold were the first—from a solution. She already knew what amino acids to code for to bind cobalt, and used directed evolution to find virus proteins that bind well to gold. She then put those two DNA sequences into the viral genome, and set them to their task. The re-engineered viruses coat themselves with the metals, then dutifully arrange themselves—like Willy Wonka's little Oompa-Loompas—into highly ordered nanowires atop a polymer, forming an electrode. Belcher chose viruses to do her bidding, though they normally would not go near inorganic materials, because they are so easy to work with. "They're only DNA and protein," she told MIT's *Technology Review*. "You don't have to worry about messing up all kinds of other metabolic processes, and you can make millions of copies in a short time."* Because they align themselves so carefully, they can pack three times more energy into a small space than the best existing batteries can manage.

By 2007 Belcher's team had grown fully functioning self-

---

* MIT *Technology Review*, September 28, 2006.

assembling batteries. Those first batteries took the form of Saran Wrap–thin sheets centimeters square, which could be built into credit cards or made into a transparent stick-on battery something like a Band-Aid. Tiny 3-micron batteries came next, which might be used in implantable medical devices. A third form, which Belcher is now actively pursuing, is thread-shaped, for integrating into textiles.

When she began her research, Belcher thought the applications for her biologically assembled batteries would be small. Now she is looking into materials for higher-voltage batteries, even possibly batteries for cars. "A year ago I would have said that was impossible," she says.

The potential uses of biology continue to multiply. At the Institute for Collaborative Biotechnologies at University of California, Santa Barbara, Daniel Morse, a professor of molecular genetics and biochemistry, is figuring out how sea creatures synthesize such remarkably strong materials (mollusk shells, coral reefs, pearls) with a precision of nanoscale fabrication that far exceeds the most modern human engineering. Of particular interest is the orange puffball sponge, which harvests silicon from seawater, using an enzyme to convert it into the spiky filaments that cover its body. In one experiment, Morse and colleagues replaced the seawater with aqueous zinc nitrate and got the sponges to deposit zinc oxide onto glass. That may prove useful for researchers at the U.S. Army Laboratories, who are putting zinc oxide on organic polymers for portable, flexible solar cells.

Mining the Sky: As windmills on the ground grow taller and their turbines bigger, a few start-ups are advocating a bigger leap still: into the fierce winds that blow steadily at high altitudes. Ken Caldeira, an atmospheric scientist at the Carnegie Institution of Washington's Department of Global Ecology, estimates that just 1 percent of the energy in the jet stream could power the whole of civilization. One company exploring that potential is Sky Wind-Power, based in Colorado, which is developing a "flying generator" that looks like a big silvery H high in the sky. Rotors at each point will spin to provide lift and to generate electricity. The electricity will be transmitted back down to earth over aluminum cables, which will also serve to tether the craft. If the wind dies, electricity will turn the rotors in reverse to keep the generator airborne. Electronic controls will automatically adjust the pitch of the rotors and the craft's elevation and "attitude," (that is, its pitch, roll, and yaw) in response to changing wind conditions. Ultimately, the generator is designed to fly at thirty-five thousand feet, where the winds are a hundred times stronger than winds close to the ground. It will need to be extraordinarily robust, able to handle not only the high winds, but also lightning, storms, and turbulence, without regular maintenance.

Sky WindPower says an array of six hundred flying generators would make 12,000 megawatts, tethered over a ground space measuring ten by twenty miles. What effect might such an array—with its six hundred long, strong tethers—have on aircraft safety? There is a precedent, the company notes: balloons tethered at altitudes up to fifteen thousand feet along the southern border of the United States—part of the Tethered Aerostat Radar System, designed to detect illegal drug smuggling.

A second company exploring high-altitude wind—Makani Power of Alameda, California—got a big boost in November 2007 when Google named it a partner in its RE<C (renewable energy cheaper than coal) initiative. Google had invested $10 million in the company a year earlier. Though Makani has been quiet about its technology, its young team includes several top kite designers and "kite-boarders"—devotees of a sport akin to windsurfing while harnessed to a kite, which adds speed and enough lift to do wild flips and leaps and spins. The best clue to the company's plans

is probably an August 2003 posting on Google's sci.energy group by Makani engineer Pete Lynn. The son of one of the world's most famous kite designers, Lynn calculated the cost and efficiency advantages of using tethered kites to replace conventional wind-mills, with "the free-flying wing replacing the rotor tip." Makani founder Saul Griffith also founded Squid Labs, an engineering company that promises "we're not a think tank, we're a do tank"; he coauthors a series of comic books called *HowToons*, which teach kids how to build gadgets.

A lower-tech design from Magenn Power in Canada uses a helium-filled blimp that looks and turns like a water wheel about a thousand feet up in the sky. The balloon can be rapidly deployed, and can operate in a wide range of wind speeds—from four miles per hour to sixty. It is intended primarily for remote locations without grid power.

Wubbo Ockels, a former astronaut (the first Dutchman ever in space) and now a professor at Delft University of Technology, has backing from Royal Dutch Shell to develop a "ladder mill": a line of remote-controlled kites attached to a big loop of cable like the chairs on a ski lift. As the kites are lifted by the wind to heights of a thousand to ten thousand feet, they pull the cable up behind them. The kites change position for the descent, reducing wind resis-tance and the energy required to make them dive. Looped around a wheel on the ground, the cable turns the wheel as the kites rise, driving a generator. A large ladder mill, says Ockels, could produce 50 megawatts, at a cost of 5 cents per kilowatt-hour. Though the idea has been around for years, the recent development of super-light, super-strong materials and small electronics capable of posi-tioning and aerodynamic control has made it possible, he says, to realize at large scale.

Yet another company with its sights set on the sky is Cool Earth Solar in Livermore, California, though its plan is to float inflatable solar concentrators rather than windmills. Each balloon looks like a small alien spacecraft (or a Caribbean steel drum) with a clear plastic dome on top. Suspended from elevated cables, a complicated truss system keeps the devices oriented to the light. The shiny, concave bottom concentrates light onto a high-quality photovoltaic cell at its center; the dome protects the cell from rain, dirt, and bugs. Founder Eric Cummings, who spent eight years at Sandia National Laboratories, says that by radically reducing both the materials and real estate requirements, Cool Earth will cut the price of solar electricity to 29 cents a watt by 2010. He raised about a million dollars in venture capital in 2007.

The most extreme proposal for mining the sky has been explored for several decades by both NASA and the Department of Energy, which together have spent some $80 million investigating the potential for launching solar collectors into outer space. Circling the earth to follow the sun through an endless day, the satellites would absorb continuous light, five times as much light as the best solar locations on the ground.

In the spring of 2007 the idea won an influential new champion: the Department of Defense. Describing the development of new energy resources as vital to the protection of U.S. safety and economic interests, the National Security Space Office Advanced Concepts Office (which calls itself "Dreamworks") launched an extraordinary open-source research effort on space-based solar power. More than a hundred experts were invited to participate in an online collaborative effort, culminating with the October presentation of an enthusiastic report. A dazzling animation prepared by the Pentagon showed orbiting solar arrays shooting big blue beams of electromagnetic energy to a receiver on earth. (The beams would be no more powerful than the noonday sun, reporters were reassured, and could not be used as weapons.) The report outlined the rather major challenges to be overcome, including the creation of "low-cost space access." But it also noted that solving those problems would facilitate all kinds of other happy developments, including space tourism, asteroid mining, and eventually "settlement to extend the human race."

Undoing the Damage: Beyond these futuristic efforts to invent whole new ways of making energy, a number of innovators are exploring ways to remove the excess carbon dioxide that has already been dumped into earth's atmosphere. The most practical, near-term way to do that is by growing more biomass: restoring the forests and prairies that absorb carbon into their own structures and soils. Some are exploring going a further step to make that carbon uptake permanent: burning the biomass for electricity or converting it to fuel, but then capturing the carbon emissions and sequestering them in the ground. They would use biomass, in other words, as a big carbon vacuum.

Like Nate Lewis, Klaus Lackner, chair of the Department of Earth and Environmental Engineering at Columbia University, thinks man-made technology can improve on nature. He proposes to build millions of "artificial trees" that would be a thousand times more efficient than real ones at taking carbon from the atmosphere. His prototype looks like a big furnace filter, with layers of ruffled leaves of permeable material coated with sodium carbonate. As the air wafts through the filter, the sodium carbonate will combine with the carbon dioxide to become sodium bicarbonate; periodically, a liquid will flush the leaves, washing the bicarbonate into solution. That solution will go to a separator, where electrodialysis will turn it back into carbon dioxide (for sequestration) and sodium carbonate (for reuse in the filter). A unit the size of a forty-foot shipping container standing on end, says Lackner, would remove a ton of carbon dioxide a day.

Though the system has similarities to those under development for coal plants, it has still bigger obstacles to overcome. To begin, air is just 0.04 percent carbon dioxide, compared to the 10 to 12 percent carbon dioxide in flue gases, which means Lackner's filters will have to process 250 times the volume of gases handled at a power plant to capture an equivalent amount of carbon dioxide. They will need to be robust enough to function remotely and without maintenance through rain, wind, sleet, and snow. Though they could be built taller, to increase surface area and take advantage of the breezier conditions farther off the ground, at the tractor-trailer size the world would need 30 million of them just to take care of the carbon dioxide emissions produced globally by the transport sector.

The biggest problem is how much electricity the filters consume: if made from fossil fuels, that energy generation will produce nearly half a ton of carbon for every ton captured by the air filters and sequestered in the ground. Over time, Lackner anticipates shifting to low-carbon energy sources; the towers themselves, he says, could serve as support towers for wind turbines to supply their own energy. The great advantages, in Lackner's view, are that his scrubbers can be put anywhere—to capture emissions that originated at tailpipes or in home furnaces or in China—and that capture towers can be located in the best places to use or bury that carbon dioxide: greenhouses growing tomatoes, oil fields that need carbon dioxide for extraction, or the best saline aquifers for sequestration.

Though a nine-foot prototype built by Tucson-based Global Research Technologies absorbs just 50 grams of carbon dioxide a day, Lackner calculates that when fully developed, his carbon dioxide–eating tree farms will be hundreds of times smaller than wind farms capable of avoiding the same amount of carbon dioxide. Champions of the technology include Jeffrey Sachs, director of the Earth Institute at Columbia, and climate scientist Wally Broecker, a geochemist at Columbia's Lamont-Doherty Earth Observatory, who is famous for his work on the role of global ocean currents (the "conveyor belt") in abrupt climate change. Early funding came from Lands' End founder Gary Comer, who between 2000 and 2006 spent more money than the National Oceanic and Atmospheric Administration on research into the dangers of abrupt climate change.

Perhaps the wildest idea of all has been proposed by Alfred Wong of UCLA, who in Chapter 7 was making an artificial aurora borealis down the road from Chena Hot Springs. The auroral oval is one of the two places where the magnetic shield of the earth opens up to outer space, allowing the charged particles of the solar wind to slip into the atmosphere: that's what causes the northern lights. Wong proposes using that opening as what the May 2007 *Economist* called a "Stairway to Heaven" for carbon dioxide.

Wong's plan is to ionize carbon dioxide (that is, give it a negative charge) by zapping dust in the atmosphere with powerful lasers to release the electrons carbon dioxide likes to bind with. Those negatively charged carbon dioxide ions would then travel toward the sky along the lines of magnetic force—which in the polar regions point almost straight up. When a charged particle is in this magnetic field, it moves in an upward spiral. Above seventy-five miles it begins to spin more slowly but converts that lost spinning motion to more rapid upward motion. Wong would zap his carbon dioxide ions with radio waves to send them into the proper spin, so they would have enough energy to journey all the way to outer space; the solar wind, he believes, would provide a powerful extra boost. (Satellites have already detected oxygen leaving earth's atmosphere through a similar mechanism.) He calculates that even if he powered his lasers and radio antennae with fossil-generated electricity, he would still eject more carbon dioxide than he would make. And his neighbor, Bernie Karl, wants to supply him with geothermal power.

The wilder the solution, of course, the more caution required.

Some ideas now circulating—for reengineering entire systems in the earth—invite such vast unforeseen consequences that they are best left alone. Fertilizing vast reaches of the ocean to produce algal blooms, for instance, could disrupt marine ecosystems worldwide.

Nuclear Fission: As a carbon-free source of energy, nuclear fission must be part of the discussion. Most of the existing nuclear capacity in the United States—currently providing 20 percent of our electricity—is due to be retired soon, meaning that either alternative carbon-free sources of energy will have to fill the gap or new nuclear plants will need to be built.

In September 2007, NRG Energy submitted the first application in twenty-nine years for an operating permit for a new nuclear power plant in the United States, to be built in South Texas. Around the world, according to the International Atomic Energy Agency, 435 reactors in thirty countries provide 15 percent of the world's energy, with another thirty-two under construction. Much of that expansion is in Asia: In February 2007, China contracted with U.S.-based Westinghouse Electric to buy four pressurized water reactors. India plans to expand its share of nuclear-generated energy eightfold to 10 percent by 2022.

In the United States, if subsidies and taxpayer-funded insurance are eliminated, nuclear power may not be able to compete with cheaper energy sources, even with a price on carbon. But that's a question the market can answer, once the playing field has been leveled.

The more difficult problems remain: how to prevent proliferation of nuclear material that can be turned into weapons, and how to safely transport and dispose of nuclear waste—questions that remain unanswered even after four decades of effort and debate. Twenty years ago, the government and scientific community settled on long-term geological storage at Yucca Mountain in southern Nevada. Today, however, Yucca Mountain is still not operational, and many scientists are increasingly doubtful that secure geologi-cal storage can be guaranteed for the one hundred thousand years the material will remain radioactive. Meanwhile, the United States stores nuclear waste on an ad hoc basis at dispersed sites that may be vulnerable to terrorist attack or theft.

In the wake of the move by NRG and other companies to revive the nuclear power industry, Republican Senator Orrin Hatch of Utah and Democratic Senator Harry Reid of Nevada advanced legislation promoting development of reactors that would use a hybrid of uranium and thorium fuel, saying it could cut high-level nuclear-waste volumes in half. Since 1992, a Virginia-based com-pany, Thorium Power, has been working on such a process, which keeps most of the fuel in the reactor for almost ten years (current reactors require new fuel every one or two years) and produces by-products that would be hard to steal because of their high levels of gamma radiation. The company plans to test the technology at a reactor in Russia by the end of the decade.

Ultimately, given the dangers of climate change, every poten-tial source of carbon-free energy must be on the table. The U.S. government should make solving the problems of nuclear waste disposal and security a national priority. As Bill McKibben, author of *Deep Economy: The Wealth of Communities and the Durable Future*, wrote in a November 16, 2006, article in the *New York Review of Books*, "We should be at least as scared of a new coal plant as of a new nuclear station. The latter carries certain obvious risks . . . while the coal plants come with the absolute guarantee that their emissions will unhinge the planet's physical systems."

FOR INNOVATORS WHO FIND SAFE and workable solu-tions to global warming, the rewards will be staggering. As a start-ing point, in February 2007, Richard Branson, head of the Virgin Group, offered a $25 million prize to whoever finds a way to remove a "significant amount" of greenhouse gases—equivalent to 1 billion tons of carbon dioxide or more—from the atmosphere every year for at least a decade. Like the $10 million "X Prize" for private spaceflight won in 2004, Branson's proposal was inspired by the £20,000 prize offered by the British Parliament in the eighteenth century for finding a way to measure longitude at sea—one of the most pressing technological challenges of its day. The scientific grandees at the time were sure that the answer lay in the stars, and they set about tracking the minute movements of the moon and amassing enormous catalogs of constellations so that a ship captain could fix his position by tracking the movement of the heavens. A better answer came from an unexpected and decidedly less prestigious direction: clockmaker John Harrison, who went about building the world's most accurate clock, one that could keep accurate time far away from shore, despite the pitch and yaw of the open ocean and the warping effects of tropical humidity. A captain with such a clock could calibrate his position simply by comparing local noon (when the sun reached its zenith over the ship's deck) with the time in Greenwich.

In his January 31, 2007, column for the *New York Times*, David Leonhardt laid out the essential advantages of such prizes: "They reward nothing but performance, and they ensure that anyone with a good idea—not just the usual experts—can take a crack at a tough problem." Citing the million dollars offered by Netflix to anyone who could devise a better way to recommend movies, which attracted 27,000 experts and amateurs from 160 countries, Leonhardt concluded that a cap on carbon would bring the same far-ranging burst of innovation to the problem of global warming. A carbon cap "would effectively create a multibillion-dollar prize—in the form of new customers—for whichever companies came up with efficient energy sources."

Some economists have proposed a carbon tax rather than a cap-

and-trade system, arguing that a tax would not only provide the price signal needed to spur changes in behavior but also generate revenues to fund research in clean energy and help low-income households pay their energy bills. A cap-and-trade system could also create government revenue, by auctioning off some of the allowances. But more important, a cap would mobilize far more money than the government ever could to solve the problem of global warming, by creating a marketplace where anyone who can reduce the levels of carbon in the atmosphere can sell those reductions for a profit. A tax creates no such market, and so fails to enlist the full range of human potential in a struggle where every bit of creativity is needed.

The most critical failing of a tax is that it requires lawmakers to guess how high that tax must be in order to drive emissions down far enough to ensure the planet's safety. The price of guessing wrong, and allowing emissions to rise past the point where the planet may begin to change in irreversible ways, is a gamble no one should be willing to take.

Some legislators have proposed creating a cap-and-trade regime but adding what they call a "safety valve" to contain the costs of compliance: anytime the market price of a ton of carbon rises above a certain level, new allowances would be issued. Though this may seem a minor modification, it in fact completely undermines the purpose of an emissions cap. First, it would deter investment, because investors would know they could never earn more than that predetermined price; when America capped the price of natural gas, new exploration and innovation ground nearly to a halt. By dampening investment in innovation, such a provision in the long run would actually make the cost of solving global warming far higher. Worst of all, flooding the market with new emissions allowances not balanced by any real reductions would defeat the ultimate objective of climate legislation: to ensure that carbon levels in the atmosphere are kept below the dangerous tipping point. An inviolate cap without an escape hatch provides the only guarantee that we will stabilize our climate.

THIS LITTLE BLUE PLANET is the only place in the universe we know of that shelters life, perhaps the only place in the universe with an atmosphere capable of doing so. We have already altered the fragile envelope that protects us. As we continue to alter it more profoundly each day, it is impossible to predict with certainty what will become of our planet and the life it contains. While some argue that uncertainty means we should wait till the picture is clearer, in fact it makes acting now even more urgent. If we mobilize now and get better-than-expected news later, then the only mistake we will have made is to have invested more in clean energy than we would have otherwise—and since we will have to develop alternatives to fossil fuels eventually, that would be money spent too early, not money misspent. If we do not act now, however, and the worst-case scenarios unfold, well then our chance for acting will have been foreclosed. As ice caps melt and seas rise, as droughts expand their reach, species go extinct, and coral reefs disappear, there will be no going back and nothing left for us to do.

To save the planet from calamity, innovation and deployment of known technologies must occur now at a pace as intense and a scope as vast as the settlement of the western frontier. (There, too, financial incentives motivated a huge and sustained individual effort.) To stabilize the global climate, the world must produce at least 14 trillion watts of carbon-free energy by 2050—about as much power as we now get from the entire fossil energy business. In Europe, where a cap-and-trade regime is already in place, companies are taking the first steps. A 2006 MIT study found that—despite the conventional wisdom that the European carbon trading system had been ineffectual, it had in fact achieved emissions reductions.★

But paralysis in Washington means that innovators here face an expanding competitive disadvantage—and that the planet remains in deep peril. As the editors of the *Economist* observed in a special issue devoted to the problem (June 2, 2007), "Business has changed nothing like enough to have a chance of averting global warming—but given the right incentives, it can. Whether that happens or not will be largely determined in America." What the editors meant, of course, is that U.S. leaders need to rise to the challenge, first by establishing a national carbon cap and then by leading the effort to create a binding international agreement to secure emissions cuts beyond U.S. borders.

In meeting the challenge of global warming, we are all in it together; either we all succeed or we all fail. Stabilizing the climate will require cutting greenhouse gas emissions around the globe—in Europe and Japan, which have already begun to do so; in tropical nations like Brazil and Indonesia trying to slow deforestation; and in the growing powerhouses, especially India and China.

But international agreement to binding limits will not happen without the United States—which is now the only developed country in the world not under a carbon cap—playing a central and catalytic role. As the nation responsible for the largest portion (nearly 30 percent) of man-made greenhouse gases already in the atmosphere, the United States has the obligation to lead. And as the richest nation on earth, and the sole superpower in economic as well as geopolitical terms, the United States has the unique ability to lead. Developing nations racing to modernize cannot be expected to act if the largest and richest emitter will not. Some argue that the United States should not make a move

---

★ "Over-Allocation or Abatement? A Preliminary Analysis of the EU Emissions Trading Scheme," by Denny Ellerman and Barbara Buchner.

until China and India do. But the United States has never followed China's lead in foreign policy—nor should it do so now. If Congress creates what will likely become the world's largest carbon market, however, and offers other nations the chance to participate in that market if they too cap and cut emissions, that will provide a powerful lure for them to join us, and bring enormous amounts of financial capital into new low-carbon investment opportunities all over the world.

The starting point must be the U.S. Congress. Mobilization on the necessary scale will occur only when U.S. leaders pass the law that will allow alternative energy sources to compete fairly with oil and coal: a hard cap on global warming pollution. The legislative limits must rely on the most rigorous definitions of carbon reductions, establish strong monitoring and verification systems, and impose severe penalties for cheating. The new law must mandate substantial short-term and long-term reductions, so that businesses not only know they need to take action, make investments, and find new technology *now*, but also have the certainty that the market for low-carbon technologies will surely and steadily improve. The magnitude of the emergency places these principles above politics, beyond ideology.

Only when legislators make it a regulatory certainty that global warming pollution will be limited will U.S. companies invest seriously in solar, biofuels, wave energy, and clean cars. DuPont CEO Chad Holliday echoes the sentiments of many of the executives, entrepreneurs, and investors you have met on these pages. He thinks about doubling the number of scientists he has working on cellulosic ethanol. But without knowing what the regulations on carbon emissions are going to be, he and DuPont's shareholders cannot evaluate the market or calculate how much to invest in research. In Holliday's words, "You need some certainty on the incentives side and on the market side, because we are talking about multiyear investments, billions of dollars that will take a long time to pay off." The CEOs who lead the World Business Council for Sustainable Development endorse that assessment: "To scale up investment flows into new low-greenhouse-gas technologies and . . . rapidly deploy those technologies across the world, policy efforts must align with long-range business investment cycles. The bulk of potential private capital will remain uncommitted until definitive policies emerge."

That caution is commensurate with the high risk attached to these investments, which is why the accelerating effect of a carbon cap is so vitally needed. The capital risk for clean energy is much higher than it was for information technology, says John Doerr. Where Google needed just $25 million of venture capital and only two years to get to positive cash flow, Bloom Energy (a fuel cell company Doerr is now backing) will require ten times that much money and five to seven years to get to positive cash flow. The technology risk is also higher, because the industry is newer and the reservoir of trained talent shallower. "There aren't that many engineers and scientists who can take what we know about thin film sputtering from semiconductor manufacturing and apply it to the particular chemistry of CIGS,* or who can design better fuel molecules and then modify bugs to eat sugar and secrete those molecules," says Doerr. "These founders require multidisciplinary talents, not just biology or chemistry but both, and also fuels-supply chain expertise. They have to pay more attention to policy, whether it's California's global warming bill or renewable standards or cap and trade." But, Doerr adds, "that also means they're going to be more nimble as we get more change. And I think this is a marathon, not a sprint."

In fact, the sheer scale of the problem is one reason our sense

---

* Photovoltaic film made of copper, indium, gallium, and selenium.

of alarm has given way to excitement and hope. The question is no longer just how to avert the catastrophic impacts of climate change, but which nations will produce—and export—the green technologies of the twenty-first century. A cap-and-trade system for carbon dioxide will mean billions of dollars for the innovators who figure out how to save the planet, and provide the opportunity to mobilize virtually every realm of economic activity.

Over the next thirty years, according to the International Energy Agency, governments and private investors will spend no less than $10 trillion to update and expand the global electricity infrastructure. How that money gets spent will be largely determined by what we demand of our political leaders today.

We have before us an extraordinary opportunity: to harness the power of the United States of America's huge and dynamic markets to ensure a safe future. None of us any longer can stand by and watch; all of us must engage as citizens to demand that our country lead the world to solve the climate crisis. Enacting a cap on carbon will gather U.S. ingenuity and resourcefulness to serve a higher purpose: protecting this planet for generations to come. We have the talent and a brief window of time to create the world of possibilities. All we need is the resolve.

ACKNOWLEDGMENTS

This book exists only because of the generous contributions of many people. We are particularly grateful to John Doerr, whose impassioned description of these new energy entrepreneurs first inspired the idea. Throughout the process, John generously shared insights gained in his many years of investing in world-changing technologies. His colleague Ellen Pao provided some of the earliest introductions to important inventors and entrepreneurs. Charles Byrd of MissionPoint Capital, Mike Danaher of Wilson Sonsini Goodrich & Rosati, and Christine Hinton and Marc van den Berg of VantagePoint Venture Partners also gave us important early help in our search.

Nick Nicholas, Ann Doerr, Jeanne Donovan Fisher, and Susan Mandel, all members of the Environmental Defense Fund board, provided early encouragement, as did our former colleague Tom Belford, consulting from his new home in New Zealand. We were also fortunate to have the guidance of three of the world's most eminent climate scientists: William L. Chameides, dean of Duke University's Nicholas School of Environmental Policy; Michael Oppenheimer, director of the program in Science, Technology and Environmental Policy at Princeton University's Woodrow Wilson

School of Public and International Affairs; and Stephen W. Pacala, director of the Princeton Environmental Institute.

Our colleagues at Environmental Defense Fund have supported and helped us in countless ways. From beginning to end, Executive Vice President David Yarnold has played many crucial roles—managing the project, marshalling the resources to tackle it, and ensuring that we did justice to the excitement of these unfolding stories. For sharing their deep knowledge of clean air history, power markets, ecosystems, cars, the Amazon, economics, policy, and science, we particularly wish to thank Jeffery Greenblatt, Nat Keohane, Mark Brownstein, Vickie Patton, Steve Schwartzman, Scott Anderson, Diane Regas, Rod Fujita, Doug Rader, and John DeCicco. For her understanding of the special insight Environmental Defense Fund could bring to this project, and her able help, Cynthia Hampton.

It has been a pleasure to work with Norton, and we thank our editor, Starling Lawrence, for demanding clarity throughout and for editing so carefully and gracefully. We also wish to thank Bill Rusin, Molly May, and Louise Brockett.

Our agent Gail Ross and her colleague Howard Yoon have stood by our side every step of the way. Merrill McLoughlin provided superb editing and complex logistics management. We also thank David Rubin, Jennifer Freeman, Brian Gallagher, Paul Katz, Melanie Janin, Sally McCartin, Mark Fortier, Joel Plagenz, Jeremy Meyers, and Karen Kenyon.

Fred thanks his wife, Laurie Devitt, and son, Jackson Krupp. Miriam thanks her husband, Charles Sabel, and daughter, Francesca, as well as Carmen Chang, Antoine Predock, and the Klein Young, Hasan, Zucker Goren, Sundaram, Hogue, Burrow, Thom, Brown, and Jurcak families.

Most of all, we thank the innovators for giving us so much time and for their efforts to bring this new world into being.

RESOURCES

In addition to the sources cited throughout the book, we are
indebted to the up-to-the-minute, insightful coverage provided by
these online resources:

alarm:clock www.thealarmclock.com

Audubon www.audubonmagazine.org

AutoblogGreen www.autobloggreen.com

Bloomberg News www.bloomberg.com/news/

Business Week www.businessweek.com

Clean Edge www.cleanedge.com

Cleantech www.cleantech.com

ClimateBiz www.climatebiz.com

Climate 411 www.environmentaldefenseblogs.org/climate411/

CNET www.news.com

EcoGeek www.ecogeek.org

GreenBiz www.greenbiz.com

Green Car Congress www.greencarcongress.com

Greentechmedia www.greentechmedia.com

Greenwire www.eenews.net/gw/

Green Wombat http://blogs.business2.com/greenwombat/

Grist http://grist.org/

MIT *Technology Review* www.technologyreview.com/

Nature www.nature.com/

New York Times www.nytimes.com

Point Carbon www.pointcarbon.com

Popular Mechanics www.popularmechanics.com

Popular Science www.popsci.com

Red Herring www.redherring.com

Renewable Energy Access www.renewableenergyaccess.com

Science www.sciencemag.org

Science News www.sciencenews.org

Solarbuzz www.solarbuzz.com

Sustainablog www.sustainablog.org

TreeHugger www.treehugger.com

VentureBeat www.venturebeat.com

Wall Street Journal www.wallstreetjournal.com

Worldchanging www.worldchanging.com

INDEX